STRUCTURAL DRAFTING

A PRACTICAL PRESENTATION OF DRAFTING AND DETAILING
METHODS USED IN DRAWING UP SPECIFICATIONS FOR
STRUCTURAL STEEL WORK

By FRANK O. DUFOUR, C. E.

ASSISTANT PROFESSOR OF STRUCTURAL ENGINEERING,
UNIVERSITY OF ILLINOIS

ILLUSTRATED

British Library Cataloguing-in-Publication Data
A catalogue record for this book is available from the
British Library

Technical Drawing and Drafting

Technical drawing, also known as 'drafting' or 'draughting', is the act and discipline of composing plans that visually communicate how something functions or is to be constructed.

It is essential for communicating ideas in industry, architecture and engineering. The need for precise communication in the preparation of a functional document distinguishes technical drawing from the expressive drawing of the visual arts. Whereas artistic drawings are subjectively interpreted, with multiply determined meanings, technical drawings generally have only one intended meaning. To make the drawings easier to understand, practitioners use familiar symbols, perspectives, units of measurement, notation systems, visual styles, and page layout. Together, such conventions constitute a visual language, and help to ensure that the drawing is unambiguous and relatively easy to understand.

There are many methods of constructing a technical drawing, and most simple among them is a sketch. A sketch is a quickly executed, freehand drawing that is not intended as a finished work. In general, sketching is a quick way to record an idea for later use, and architects sketches in particular (in a very similar manner to fine artists) serve as a way to try out different ideas and establish a composition before undertaking more finished work. Architects drawings can also be used to convince clients of the merits of a design, to enable a building constructer to use them, and as a record of completed work. In a similar manner to engineering (and all other technical drawings), there is a set of conventions (i.e particular views, measurements, scales, and cross-referencing) that are utilised.

As opposed to free-sketching, technical drawings usually utilise various manuals and instruments. The basic drafting procedure is to place a piece of paper (or other material) on a smooth surface with right-angle corners and straight sides – typically a drawing board. A sliding straightedge known as a 'T-square' is then placed on one of the sides, allowing it to be slid across the side of the table, and over the surface of the paper. Parallel lines can be drawn simply by moving the T-square and running a pencil along the edge, as well as holding devices such as set squares or triangles. Other tools can be used to draw curves and circles, and primary among these are the compasses, used for drawing simple arcs and circles. Drafting templates are also utilised in cases where the drafter has to create recurring objects in a drawing – a massive time-saving development.

This basic drafting system requires an accurate table and constant attention to the positioning of the tools. A common error is to allow the triangles to push the top of the T-square down slightly, thereby throwing off all the angles. Even tasks as simple as drawing two angled lines meeting at a point require a number of moves of the T-square and triangles, and in general drafting this can be a time consuming process. In addition to the mastery of the mechanics of drawing lines, arcs, circles (and text) onto a piece of paper – the drafting effort requires a thorough understanding of geometry, trigonometry and spatial comprehension. In all cases, it demands precision and accuracy, and attention to detail.

Conventionally, drawings were made in ink on paper or a similar material, and any copies required had to be laboriously made by hand. The twentieth century saw a shift to drawing on tracing paper, so that mechanical copies could be run off efficiently. This was a substantial development in the drafting process – only eclipsed in the twenty-first century with 'computer-aided-drawing' systems (CAD). Although classical draftsmen and women are still in high demand, the mechanics of the drafting task have largely been automated and accelerated through the use of such systems. The development of the computer had a major impact on the methods used to design and create technical drawings, making manual drawing almost obsolete, and opening up new possibilities of form using organic shapes and complex geometry.

Today, there are two types of computer-aided design systems used for the production of technical drawings; two dimensions ('2D') and three dimensions ('3D'). 2D CAD systems such as AutoCAD or MicroStation have largely replaced the paper drawing discipline. Lines, circles, arcs and curves are all created within the software. It is down to the technical drawing skill of the user to produce the drawing – though this method does allow for the making of numerous revisions, and modifications of original designs. 3D CAD systems such as Autodesk Inventor or SolidWorks first produce the geometry of the part, and the technical drawing comes from user defined views of the part. This means there is little scope for error once the parameters have been set.

Buildings, Aircraft, ships and cars are now all modelled, assembled and checked in 3D before technical drawings are released for manufacture.

Technical drawing is a skill that is essential for so many industries and endeavours, allowing complex ideas and designs to become reality. It is hoped that the current reader enjoys this book on the subject.

CONTENTS

WOOLWORTH BUILDING IN PROCESS OF CONSTRUCTION
Cass Gilbert, Architect
Courtesy of Thompson-Starrett Company, New York City

INTRODUCTION

A STEEL skyscraper, like the fifty-five story Woolworth Building in New York, is an architectural and engineering triumph and its erection is a never-ending source of interest and wonder to the spectator. Throngs of interested persons are always standing near while the giant derricks raise the columns and girders aloft, swing them into place, and leave them to be fastened securely by the men with white hot rivets and pneumatic hammers. And yet in all the confusion of noise and bustle of the workmen, the control of the engineer with the blue detail sheets before him is in evidence. Every piece of steel has a certain place in that great structure and, although made miles from the scene and possibly in different mills, every piece fits exactly down to the rivet holes. The skeleton grows before your eyes, a potent illustration of the might of minds, of the value of accuracy of detail and organization of mechanical forces.

¶ Do we not too often, in the midst of our wonder and appreciation of this marvelous work of man, forget from whence all this order and system sprang? Do we not lose sight of the careful calculation of the size and thickness of every angle, plate, and girder, the location of the holes, and the number and size of the rivets to be used? When we think also of the checking and rechecking of all the calculated results in order to avoid mistakes in dimensions and in order to make sure that no structural member is to be called upon to bear strains beyond its strength, we begin to see how it is possible for a whole building to be put together without one alteration, without one single piece being sent back to the mill. The man who stands boldly on a swinging girder and looks down from dizzy heights to the throng below is a very unimportant man compared to the designer, the detail man, the checker, and the steel mills' superintendent, who carry the work safely and accurately to a finish.

¶ The author in this article has spoken from a wide experience in this line of work and his practical instructions regarding drafting methods, systems of costs, record sheets, and the detailing of the elements of structural steel will be found of exceptional value to every trained and untrained man. It is the hope of the publishers that the work will be of service to a wide circle of students and general readers.

HOTEL LASALLE, CHICAGO
Holabird and Roche, Chicago, Architects. George A. Fuller Company, Contractors, Chicago

JUNE 5 1909
B4024

BLACKSTONE HOTEL, CHICAGO, IN PROCESS OF CONSTRUCTION
This Fireproof Hotel Is of Steel and Tile Construction Throughout, with Brick Exterior.
Marshall & Fox, Chicago, Architects.   Geo. A. Fuller Company, General Contractors

STRUCTURAL DRAFTING

PART I

DRAFTING ROOM EQUIPMENT AND PRACTICE

Introduction. Structural drafting may be defined as the art of making drawings of certain objects and placing thereon dimensions and other notes which when taken together will convey the necessary information for the manufacture and in some cases the erection of the structure under consideration.

In the making of these drawings great accuracy in drafting is not necessarily required. The chief requisites are that the lettering and dimensions should be so clear that no misunderstanding is possible. Dimensions not given should never be scaled by the draftsman or workman, but the actual value should be ascertained by consulting some one familiar with the work.

Classification of Drawings. The classes of drawings which are made in a structural drafting room are: the stress sheet; the assembly, or general detailed, drawings; and the shop drawings, or, as they are more often called, the detailed drawings.

The stress sheet is a tracing upon which is usually shown a skeleton outline of the structure upon the lines of which are marked the stresses which are caused by the traffic or other forces to which the structure is subjected, and also the size and shape of the member designed to withstand these stresses.

The assembly or general detailed drawings usually give several views of the structure as it appears after it has been erected. On these views are shown to scale the members as they appear in the finished structure together with all the rivets and other details necessary for its completion. The overall dimensions are usually given and also any other dimensions which are necessary for the draftsman to complete the shop drawings. While the size of the members and their connections, as well as the number of rivets required, are always given, yet in a few cases the length of the member or shape and the individual spacing of the rivets are also given.

The shop drawings, or detailed drawings as they are more often called, consist of views of a certain member of the finished structure so dimensioned that it may be constructed by the men in the shop. It requires greater skill and more experience to make the assembly drawings than it does the detailed drawings, but in each case the men must be familiar not only with the drafting practice but also with that of the mill and the shop.

Drafting=Room Personnel. A drafting-room force consists of an engineer, a chief draftsman, squad boss, checkers, draftsmen, and tracers.

The *engineer* has charge of the plant as well as of the drafting room and is directly responsible for the ordering of all material, the manufacturing of the structure and its shipping to the place of erection. He conducts the correspondence, keeps track of the work in the drafting room and in the shop, and, in case his plant is one of many of a large corporation, makes weekly or monthly reports to his superior officers. In case his plant is a small one, the engineer usually does most of the work of designing and estimating.

The *chief draftsman* is directly responsible to the engineer for the getting out of the detailed plans or shop drawings and also ordering of the material.

The *squad boss* reports to the chief draftsman and his duty is to keep track of and to get out the drawings of any particular structure which is assigned to him by the chief draftsman. The squad bosses usually have from three to four to as many as twenty *draftsmen* under them, according to the magnitude or the number of structures which they are responsible for.

In addition to the draftsmen are the *checkers*, certain men usually of great experience in matters relative to mill and shop as well as drafting-room practice. It is the duty of these checkers to go over the draftsmen's work, see that all errors are corrected, and then finally sign it as approved. The checker only is held responsible for mistakes which then may be left upon the sheet.

The *tracers* are for the most part young college graduates or apprentices, and their office is simply to trace the drawings which are handed to them by the draftsmen.

A fireproof vault is always a part of the equipment of every well-equipped drafting office. In it are kept the notebooks in which the computations necessary for the design and detailing of the structures are kept, and also the tracings which have been made in the drafting room. In case the drawing of any particular structure is required, the tracing is taken out of the vault, blue prints are made, and the tracing returned as soon as possible. The vault should be so equipped that whenever the door is opened the interior becomes lighted. Aside from the mechanical convenience of this arrangement, it avoids the possibility of any person being accidentally locked in, since the rule is that in case of fire the vault should be immediately closed by the one nearest to it.

Assignment of Work. When the engineer of the plant received a stress sheet from his head officer or from the designing department in his own work, he hands it to the chief draftsman. The chief draftsman makes a record of it and gives it to the squad boss who is most accustomed to that class of work. The squad boss in turn hands it to the checker or checkers and these men make details for the various parts of the structure and make layouts for the various joints. The engineer now orders the material which will be required to build the structure or assigns a checker to do so and then returns the stress sheet to the squad boss who assigns certain draftsmen to prepare the shop drawings for the structure. Draftsmen make the drawings and turn them over to the tracers to trace them.

After the tracer has finished the tracings of the sheets, he passes them to the checker who in the first place made out the details and layout and ordered the material. The checker goes over these tracings very carefully and sees that all dimensions are correct, that all material used is that which he ordered, and if the drawings are correct he signs his name to the sheet. If the dimensions or any other matter upon the drawing is found to be incorrect, the checker places a ring around it with his blue pencil which is used in checking and off to one side places the correct value. After all the apparent errors have been corrected in this manner, a consultation between the checker and the draftsman who made the drawing is held. The errors are pointed out to the draftsman who in turn checks the work to prove the checker's results. The draftsman then takes the drawing and makes the necessary changes and returns it to the checker.

Great care should be taken in making the changes that no dimensions or other notations written upon the drawing by the checker are rubbed off. The checker then examines the drawing carefully to see that all the errors which he has pointed out have been corrected. He then cleans the tracing, signs his name to it, and returns it to the squad boss. The squad boss in turn has the necessary blue prints made and turns the tracing together with the prints over to the chief draftsman, who in turn files the tracing in its proper place and gives the blue prints to the engineer of the plant who sees that they are distributed to the foremen of the various shops where they are required.

Records. A job is known by the order number which is given it when it comes into the hands of the engineer of the plant. This order number should go on all papers upon which anything concerning that structure is placed. Failure to do this will result in great confusion and much time will be lost. The penalty for persistent failure to comply with this very important method of procedure is usually dismissal.

Since the draftsman, or in fact any of the office force, may work upon more than one order during the day or week, and since it is important that the cost of the drafting or engineering work for any particular order should be known, it is essential that the men keep time cards upon which the order and the time placed upon that order is noted. Usually fractions of an hour less than one-fourth are not reported. Fig. 1 shows one of these time cards upon which is noted the work of a checker for one week. It shows that he has worked upon several orders and also shows the exact amount of time he has placed upon each one and also the rate per hour which he received. In this way it is possible to obtain the cost of engineering of any particular order when it has finally been finished.

An orderly record of the passage of the work from the time the stress sheet enters the engineer's office until the material has been shipped, and also a record of the progress of the work during erection, should be kept. This is usually kept on 3×5 cards in the engineer's office. In addition to this card-index record, a monthly report in blue print form is kept showing the progress of the various orders. For instance, the progress report would contain such items as these: Order received, layouts made, material ordered, detailed sheets

FORM O & E 153-20M-10-22-03

ENGINEERING DEPARTMENT

NAME *J. A. Frost* RATE *60*

TIME CARD FOR WEEK ENDING *August 13* 190*4*

ORDER Number	Div.	Sun.	Mon.	Tue.	Wed.	Thur.	Fri.	Sat.	Sun.	Mon.	Tue.	Wed.	Thur.	Fri.	Sat.	TOTAL HOURS	COST
B412c			6	4	5½	4	6									25½	
B413c			5½	4½	6	4	5½									25½	
Estimating																	
General																	
Holiday																	
Total			11½	8½	11½	8	11½									51	
Sick								4								4	
Vacation																	
Out																	
Total			11½	8½	11½	8	11½	4								55	

HOURS WORKED *51*

" ALLOWED (not worked)

" PAID FOR *51*

D. S. James
Engr.

Fig. 1. Draftsman's Time Card Showing Hours Spent on Order Indicated

finished, shop bills made, templet work finished, work fabricated, work shipped; and in addition to this progress report, which is made out in the office, is the report of the erector on a job in the

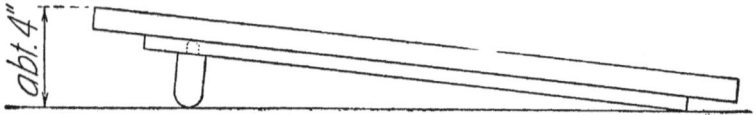

Fig. 2.   Side View of Drawing Board, Having Elevating Pegs

field.   The erector's form of report contains such headings as tend to indicate the progress in the false work; the erection of the trusses and floor system; and the amount of field riveting and painting completed.

Drafting Materials.   *Instruments.*   The drafting instruments required are: A drawing board, **T**-square, triangles of various kinds as noted below, pencils, scales, erasers and erasing shields, a set of drawing instruments, a large linen cover, and half sleeves.

The drawing board should be made of soft pine with battens upon the back in order to prevent the warping of the board.   Since few drawings in structural engineering are larger than 24×36 inches, it is not necessary to have the drafting board larger than 26×38

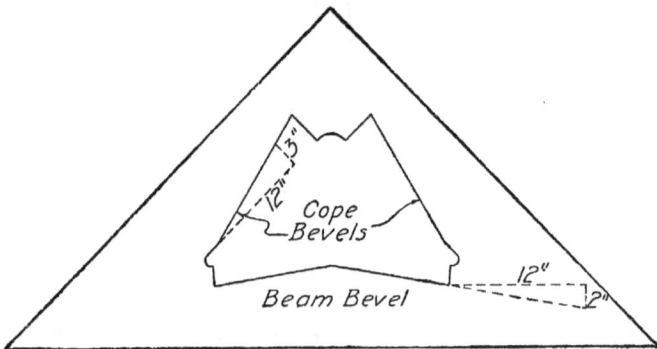

Fig. 3.   45° Triangle with Cope and Beam Bevels

inches.   A drafting board should not lie close to a table, but should be raised from the table by small legs placed at its upper edge as indicated in Fig. 2.

The **T**-square should be about 40 inches in length and should be of good quality with an amber edge upon each side.   The amber edge is of great advantage since it will allow the draftsman to see lines below that one which he is drawing and, therefore, prevent him from overrunning by drawing one line past its limiting point. Such a **T**-square may be procured for about $2.25.

The triangles should be of amber or celluloid, and should consist of the following: One 45-degree triangle with 10- or 12-inch sides; one smaller, say with 6-inch sides; two 60-degree triangles with 10-inch sides; and two with 4-inch sides. One or more of these triangles should have the beam and coping bevels fixed upon it as in Fig. 3; this will have to be done by the draftsman, since no such triangles are on the market.

The pencils used by the draftsman should be such as will make clear and black lines upon paper in case the drawing is to be traced. If the drawing is not to be traced, a harder pencil will suffice. In case a drawing is made directly upon tracing cloth, a soft pencil should be used and it should be kept sharpened. This will necessitate frequent rubbing over the sand paper pad which every draftsman should have close at hand in order to keep a good point upon his pencil. The pencil recommended for detailing where a tracing is to be made is "Koh-I-Noor, 3 H," although some draftsmen prefer 4 H or 5 H. The latter are, in the writer's opinion, to be recommended for detailing where a tracing is not required from the original. In case drafting is done directly upon tracing cloth, a 2 H pencil is the correct one to use.

A red pencil should be kept for marking upon blue prints and a blue pencil for making checks on tracings. Never use a red pencil upon tracing cloth, since it will not be easy to erase, whereas the blue-pencil mark may be washed off with gasoline or erased with a pencil eraser.

The scales required are the architect's and the engineer's. The former has certain divisions upon it and each of these divisions is divided into twelve parts which indicate inches, and these parts are in turn divided into halves or quarters or other small divisions denoting the fraction of the inch. The architect's scale which best serves the purpose is the one which has the 2-inch, $1\frac{1}{2}$-inch, 1-inch $\frac{1}{2}$-inch, $\frac{3}{4}$-inch, $\frac{3}{8}$-inch, $\frac{1}{4}$-inch, $\frac{1}{8}$-inch, $\frac{1}{16}$-inch, and $\frac{3}{32}$-inch scale. A special scale for the making of drawings to a large size or for the making of layouts is a great convenience. Such a scale is on the market and is divided so that half of an inch is equal to one inch. This scale should be in the outfit of all checkers. The engineer's scale is one on which the inches are divided into certain decimal divisions. The best scale for this is that which has its edges divided into 10, 20, 40, 50, and 60 parts of an inch. This scale is of use only in laying off bevels and natural functions of angles or in drawing outlines upon which details will be constructed with the use of the architect's scales. The tendency of young engineers to use the engineer's scale, allowing a certain decimal to equal a certain fraction of an inch, is to be discouraged because of the liability of error, and a severe penalty imposed for a second offense. Care should be taken in the use of scales such as the architect's which have different scales on the same edge in order not to get the feet which belong to the wrong scale.

A small paper clamp should be attached to the scale a short distance from the center opposite the end where the scale which the

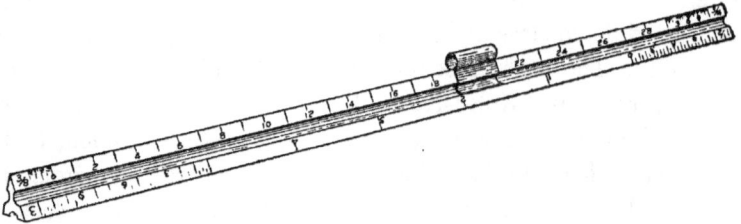

Fig. 4.   Triangular Boxwood Scale, with Scale Guard or Clamp in Position

draftsman is using is situated.   This will prevent the scale from being turned over, hence avoiding any other scale, but the one the draftsman is using at the time, turning up.   Also when the draftsman picks up the scale by the paper clamp, the end on which the scale he is using is situated will tilt downward and at once indicate to him which end he should place in position to measure what he wishes.   Fig. 4 shows one of these clamps in position for the scale as indicated.

A good ink eraser, together with a metal sheet called an eraser shield in which are various shaped holes, is an indispensable adjunct of the draftsman.   In all cases where it is necessary to erase, the ink eraser and the shield should be used.   Never use a knife to erase either upon a paper or upon a tracing cloth, for no matter how sharp the knife is, the sheet will be rubbed and ink will not run smoothly upon the place so worked over. A good soft rubber may be used for erasing pencil marks upon either the paper or the tracing cloth, although benzine, turpentine, or gasoline is much better for erasing pencil marks and cleaning off other dirty spots on the tracing cloth. Care should be taken to investigate the status of the insurance and fire laws on this point, since in many cases it is not allowable to use such inflammable materials in houses of the character of the drafting office.

An expensive set of instruments is not necessary in order to do good drafting. A good pen, a bow pen, a pair of dividers, and a compass with pencil and pen point, are all that are necessary. In many cases it is advisable to have two or more pens, one of which should be quite large, one medium, and one rather small.

Many good drafting inks are sold in the open market, and it is no longer necessary for the draftsman to make his own ink by combining India ink with water. In fact this is a distinct disadvantage, since many of the drafting inks on the market are waterproof and while tracings should not be placed so as to become wet, nevertheless it is quite an advantage to use waterproof ink upon them, so that in case they should be accidentally wetted, it will not injure them so that they can not be used.

A sheet of cambric of dark color the size of the drafting board or better still the size of the entire table and drafting board should be used to cover up the work when no one is working, since dust accumulates very readily upon the drafting board and produces much undesirable dirt and, therefore, a very dirty drawing. It is also advisable upon starting work in the morning to brush off the desk and drawing board and to wipe off the T-square and triangles with a cloth. This will prevent dirty marks appearing on the drawing when they are first placed upon them.

Detail Paper. Detail paper is the paper upon which a drawing is made before it is traced or upon which drawings are made to be used by the detailers in making up the details of the structure. Detail papers should be of buff color in order to prevent the showing of dirt upon them too easily, and also to be restful to the eye, and they should present a surface which will take a pencil or ink mark equally well, and they should not be so thin that they will not stand a great amount of erasing.

Many good papers may be bought in the open market. They may be purchased in sheets of a desired size or they may be purchased in rolls of a certain weight, and any width. When sold in sheet form they are usually sold by number of sheets; when sold in roll form, by weight. An inspection of the trade catalogue or letters of inquiry to any of the manufacturing concerns will bring further information if desired.

The standard size of a detailed sheet is 24×36 inches. Inside of this are drawn two borders each ½ inch from the other. In the

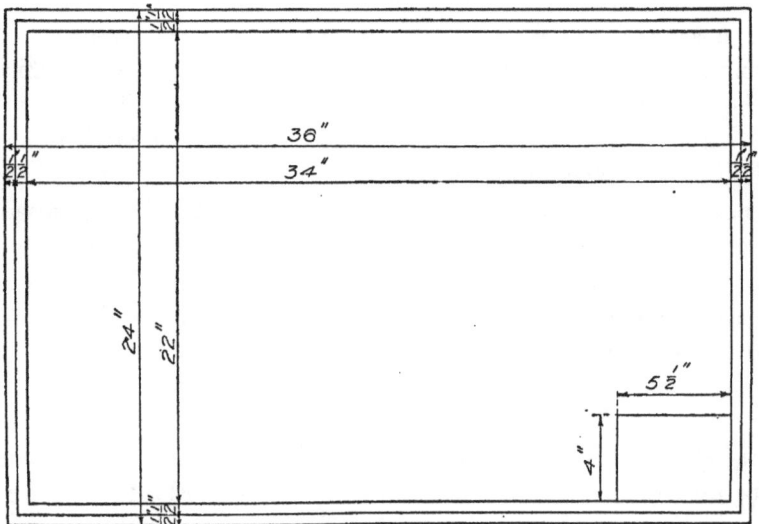

Fig. 5. Standard Detail Sheet with Dimensions

lower right-hand corner is the place for the title. The size of this block is 4×5½ inches. The 24×36-inch size is the outside dimension of the brown paper or detailed sheet. The 23×35 inch, which is the size of the first border line, is the line upon which the blue prints are cut. The second border line is the real border line of the drawing, and remains upon the blue print. Fig. 5 indicates the dimensions indicated above.

Tracing Cloth. Tracing cloth is used on account of the fact that the prints may be made from it and, therefore, any number of duplicate copies may be made available for distribution to the various departments. The drawings should be made upon the rough side of the cloth since this takes the pencil mark and also the ink better than does the smooth side. The rough side is also of a great advantage when it comes to reproducing the figure by photography. In order to make the ink take readily upon the tracing cloth and flow easily, the cloth should first be cleaned by rubbing over with powdered chalk or wiping it off with gasoline. This removes all trace of grease. Before placing the tracing cloth upon the drawing, the pink border or edge which appears upon the cloth should be torn off. If this is not done the sheet will be affected unevenly by changes in temperature, and dampness will cause the cloth to wrinkle up on account of the fact that the border is not affected by dampness and the remainder of the cloth is. This will make it difficult for the draftsman to complete his drawing in good form in case he has let it lay over for a considerable time, since the lines which he made at first will be moved from their original position by the wrinkled condition of the sheet. In case it has been forgotten to tear off this border and the sheet becomes wrinkled it is advisable to tear off the border and leave the sheet until it becomes straightened out before further drafting is done.

In some cases tracing paper is used in making small unimportant drawings. This paper should be of good quality in order that it may stand erasing, since mistakes are liable to occur and these must necessarily be corrected. The best tracing paper is brittle and will not stand much handling. For this reason its use for expensive drawings is not to be recommended.

Stress Sheet. The stress sheets for various structures are usually not made in the drafting room, but are made in the designing room of the company. Much data and many computations are made by the designer which would be of use to the draftsman in detailing. All of this information should be placed upon the stress sheets. The making of a stress sheet should be and usually is done by men of considerable experience. Plates I, II, V, and VI show stress sheets of a truss bridge, roof truss, and a plate girder, respectively, and while these can not be said to be perfect, yet they indicate the engineering practice of our larger bridge corporations and may be taken for examples. (For Plates, see pages 63, 82, 88, and 89.)

ORDERING OF MATERIAL

Since the ordering of material is of great importance it will
be discussed here somewhat at length. Although this is usually
done by men of considerable experience, yet it is advisable that the
draftsman should know the method of procedure in order that he
may be able to make the detail drawings more advisedly.

Layout. *Typical Case.* As has been mentioned before, the
checker makes details to a large size scale from which he determines

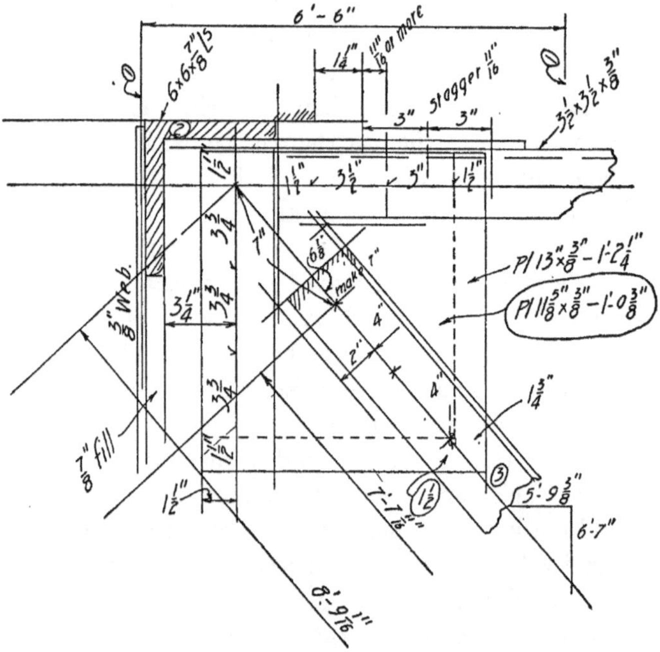

Fig. 6. Layout for Detail in Cross-Frame Connection

the size and amount of material required for certain members. In
order to illustrate this, let it be required to determine the size of
the plates and the length of the angles used in the cross frame of
the plate girder shown on Plate VI. Here the checker first lays off
the center to center of the girder to a small scale, say, 3″ to 1′. These
lines are marked 1 in Fig. 6. He next draws a web to the proper
thickness and then follows in turn the flange angles and the bracing
angles as indicated by the numbers 2, 3, etc., on the figure. The
number of rivets in the top and bottom angle and in the diagonal

should be given on the stress sheet. These should be laid off on the "layout," which is the name for the drawing that has just been made, any spacing, preferably 3 inches, being used so that the plate may be kept as small as possible. It is as a usual thing not possible to have the rivet spacing in both the diagonal and the top and the side angles of equal spacing. The number of rivets is usually put in the diagonal at about a 3-inch spacing, and the spacing of the rivets in the top and the side angles is so varied as to fill out the plate as indicated. No rivets should come closer to the edge of the plate than $1\frac{1}{2}$ inches nor further from the edge than 2 inches, and no plate should be less than an even number of inches in width although its length may be in feet, or in inches to an eighth of an inch. It is not policy to place the length of the plate in sixteenths of an inch, since the shopmen are unable to cut that close. Therefore, in determining the size of the plate the rivets should be so placed that a sufficient number should go in and the size of the plate be kept an even number of inches in width. If the rivets alone governed the size of the plate, it would be as indicated by the dotted lines in Fig. 6, and the dimensions would then be as indicated by the dimensions with a line drawn around it. The correct size of the plate is as indicated by the full line.

The length of the line from intersection to intersection point is $8'\text{-}9\frac{1}{16}''$ as indicated upon the drawing. In order to have the length of the diagonal to come out the nearest sixteenth of an inch, the distance of the first rivet from the intersection is taken arbitrarily and is as indicated here, 7 inches. It is not necessary to give this dimension to a thirty-second of an inch, since if the diagonal varies that much from the computed length, it can be drawn up into place by using a drift pin and can be riveted up without injuring the material.

Use by Checker and Draftsmen. The checker has now determined the size of the plate and the length of the diagonal angle and he records them upon the material bill which is to be sent to the mills as an order for material. This layout together with a copy of the material bill should be given to the draftsman when he starts to detail the girder. He will then have the size of a plate and the size of an angle for that particular girder so that the material which has been ordered, probably months before, and has arrived before

TABLE I

Allowances for Single Lengths

Description of Material or Rule	Allowance Inches
Web plates when ends are planed	Add $\frac{5}{8}$
Web plates when one end only is planed	" $\frac{1}{2}$
Web plates over 24″ wide, ends not planed	" $\frac{3}{8}$
Web plates under 24″ wide	" $\frac{5}{8}$
Cover plates and all other plates that must be full length when in work	" $\frac{5}{8}$
All angles where full length must be maintained	" $\frac{5}{8}$
All channels when ends are planed	" $\frac{5}{8}$
All channels when ends are not planed	" 0
All I-beams when ends are planed	" $\frac{5}{8}$
All I-beams when ends are not planed	" 0
All Z-bars when ends are planed	" $\frac{5}{8}$
All Z-bars when ends are not planed	" 0
All plates over $\frac{7}{8}$″ thick (except when ends must be planed)	" 0
Order width of all sheared plates $\frac{1}{4}$″ greater than finished width when planed edges are specified	
Order all end connection angles which must be planed or faced $\frac{1}{16}$″ thicker than specified thickness	
Order sole plates planed one side $\frac{1}{16}$″ thicker than specified	
Order sole plates planed both sides $\frac{1}{8}$″ thicker than specified	
Order Tees when ends are not planed	Add 0

the draftsman starts the detail, can be used and will be used in that girder. In case the draftsman details the cross frame without consulting the layouts and bills of material, he is liable to draw up a detail which will demand a plate larger or smaller than that ordered for that particular plate; in the first case a new plate will be required, the ordered plate being placed in the stock pile until some other job comes up in which it can be used; and in the second case the plate ordered will have to be cut down to the size of plate the draftsman has used, thus necessitating extra expense and loss of material.

In accordance with the method above stated, layouts are made, then material ordered for all details, and these layouts and copies of material bills are laid aside to be placed in the hands of the draftsman who detailed the subject. Before the material is ordered from the mills, these bills should of course be checked by another checker or by the squad boss.

In making layouts where angles are placed so that one of their legs is vertical, care should be taken to see that the horizontal leg is at the top in all cases where the angle is exposed to the action of rain and snow. If it is not in this position the angle, in case it is on a slant, will serve as a little trough down which the rain and melted snow will run into the joint at the lower end. In case the angle is not on a slant it forms a pocket-like arrangement so that the snow and ice may lodge upon it to a greater extent than if it had the vertical leg downward. Rust will result and the angle will, therefore, deteriorate. In cases such as lower chords and diagonals of roof trusses, the vertical leg of the angle should extend upward, since here the angles are not exposed to the elements and it is somewhat of an advantage that the angle should catch any dust which falls upon it, and should hold it in order to keep it from dropping to the floor beneath.

Allowances for Planing and Cutting. *Single Lengths.* When material is ordered it should be so ordered that it will be sure to be of the correct length when it gets to the shop. If the material is ordered in single lengths, that is, the length ordered to go into the finished structure without being cut in two or more pieces after it gets to the shop, it is customary to make some allowance for planing off the ends or for chance errors in the mills where the men may not be careful enough in cutting and may accidentally make the cut a short distance on one side or the other of the mark which would give the exact length The customary allowances for single lengths are given in Table I.

Multiple Lengths. In cases where there are several pieces of the same size and length, they may, for convenience in handling, be ordered in one piece at the mills and cut into lengths after they reach the shop. In this case, however, care must be taken that the multiple length is not too long to ship on an ordinary freight car. The allowances to be made in such cases and the general rules are given in Table II.

Allowances for Pin Material. In case material is ordered for pins, it is necessary that certain allowances be made for turning and for ordering in multiple. The following very general rules are given in Table III.

TABLE II

Allowances for Multiple Lengths

No.	Rule
1	No pieces more than 7 ft. long are to be ordered in multiple lengths unless under special instructions
2	In arranging multiple lengths make lengths about 30 ft. and never exceed 32 ft.
3	Never order plates over 24″ wide in multiple lengths
4	Never order plates $\frac{7}{8}$″ thick in multiple lengths
5	Never order channels in multiples unless specially instructed
6	Never order I-beams in multiples unless specially instructed
7	Never order Z-bars in multiples unless specially instructed
8	Plates and shapes to be sheared to length without finishing, add 1″ to product of length times number required
9	When planed ends are required add specified amount to each piece multiplied and add 1″ to multiple lengths so found
10	Stiffeners with fillers, add $\frac{1}{4}$″ to net length of each for planing and 1″ to multiple length so found
11	Stiffeners when crimped, order same as b-b of girder angles plus $\frac{1}{4}$″ for planing and add 1″ to multiple length so found
12	When 4 or less shapes not over 3 ft. long are ordered in multiple lengths, add $\frac{5}{8}$″ to multiple and add for planing when required.
13	When ordering fillers, allow $\frac{1}{8}$″ clearance at ends when necessary, and add for multiple as for plates
14	Make all multiples end with nearest $\frac{1}{4}$″
15	Tees under 7 ft. long may be ordered in multiple lengths. Add 2″ to length times number required and make longest multiple 24 ft.
16	If I-beams or channels are cut from long lengths allow loss of $3\frac{1}{2}$″ for each cut
17	7″×$3\frac{1}{2}$″ angles can be multiplied up to and including $\frac{3}{4}$″ in thickness

Allowances for Bending. In all cases where angles have to be bent, additional material is required. In such cases the following rules are applicable:

(1) In the case of Fig. 7a. Figure length on c.g. line of angles and add 1″ for each bend when the angle of bend is not more than about 30°; add 2″ for each bend when the angle is between 30° and 60°; over 60° ask for special instructions from the forge shop.

(2) In the case of Figs. 7b and 7c. In the case of sharply curved end angles or when sharp bends are made near ends, add to the length figured on the c.g. line as follows: 3-inch angles add 4″; 4-inch angles add 5″; 5-inch angles add 6″; 6-inch angles add 7″; 7-inch angles add 8″; and 8-inch angles add 9″.

TABLE III

Allowances for Pin Material

No.	Rule
1	Pins up to and including 4″ in diameter, add ⅛″ to finished diameter for turning
2	Pins 4″ to 6″ in diameter, add ¼″ to finished diameter for turning
3	Pins over 6″ in diameter, add ⅜″ to finished diameter and order them rough turned unless specially instructed to the contrary
4	Pins up to and including 6″ in diameter shall be ordered in multiple length of about 12 ft. Add $\frac{7}{16}$″ for each tool cut and 1″ to multiple length thus found
5	Pins over 6″ in diameter shall be ordered in single pieces and to exact length required
6	When pins are over 4″ diameter, ordered diameters must end in no fractions smaller than quarter inches

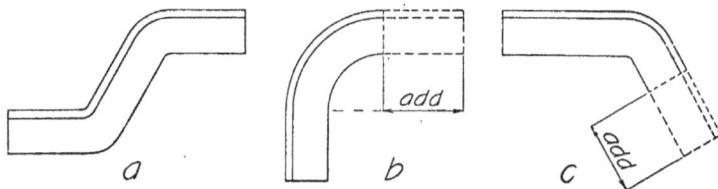

Fig. 7.   Illustration Showing Angle Bends

Shop Bills. In order to facilitate the getting out of certain articles which are of the same general form but of different dimensions, and for convenience in tabulating information relative to certain material either before or after it has been assembled into members for structures, certain bills called "shop" bills are used. These bills, which save much drafting and much letter writing, may be of almost any character to suit the practice of the plant. Figs. 8 to 26 give the headings of various bills and Fig. 27 gives the heading of a bill which is used in case it becomes desirable to change an order which has been sent in. The lower part of Fig. 27 is suitable for all of the other bills.

These bills are made on thin paper so that prints may be made from them and sent to the various shops concerned. A copy of each should also be filed in the engineer's office, and all bills of each job should be kept together by binding in some way.

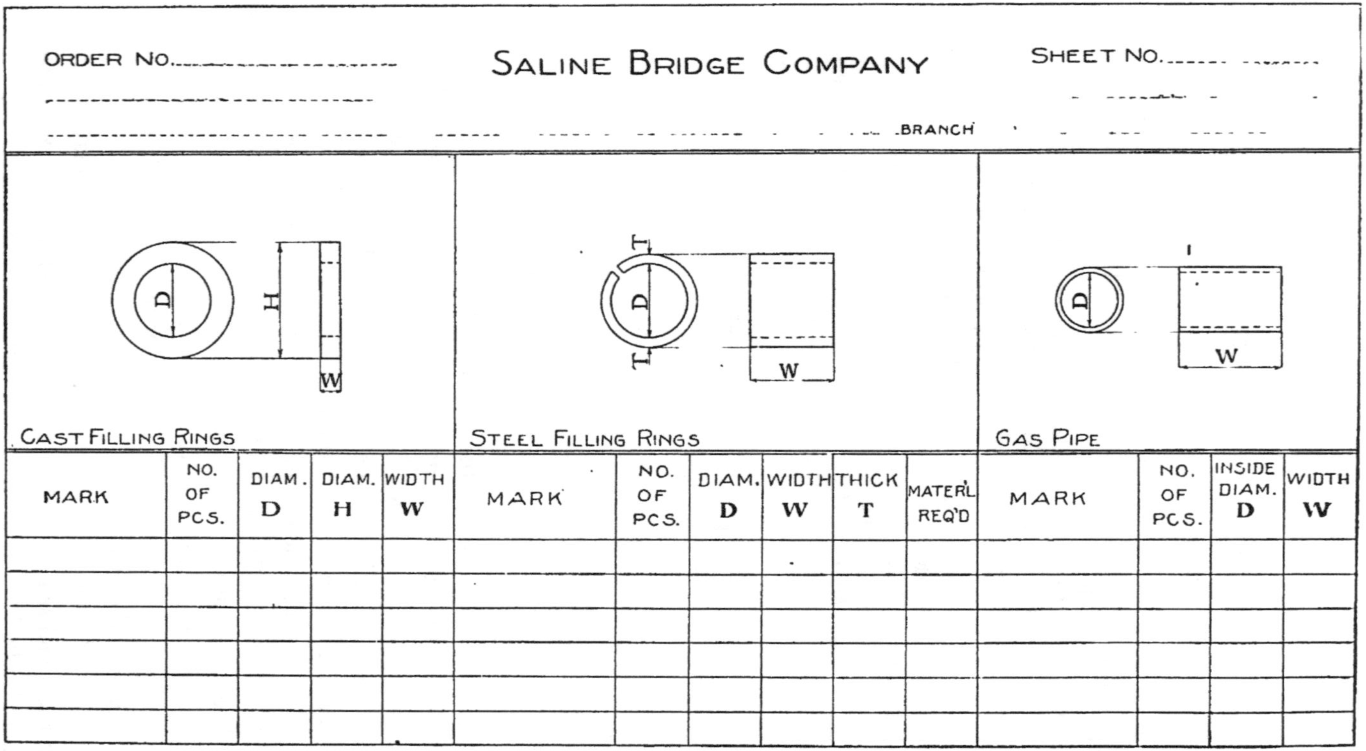

ORDER NO.________ SALINE BRIDGE COMPANY SHEET NO.________

________ BRANCH ________

CAST FILLING RINGS					STEEL FILLING RINGS						GAS PIPE			
MARK	NO. OF PCS.	DIAM. D	DIAM. H	WIDTH W	MARK	NO. OF PCS.	DIAM. D	WIDTH W	THICK T	MATER'L REQ'D	MARK	NO. OF PCS.	INSIDE DIAM. D	WIDTH W

Fig. 8. Shop Bill for Standard Filling Rings

SALINE BRIDGE COMPANY

ORDER NO._____

BRANCH_____

SHEET NO._____

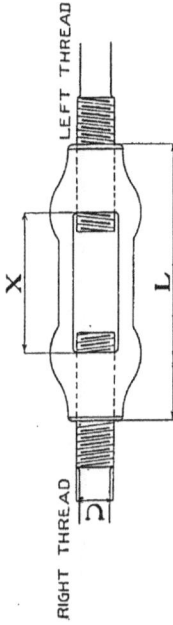

LEFT THREAD

RIGHT THREAD

X

L

LEFT THREAD

RIGHT THREAD

L

SLEEVE NUTS

MARK	NO. OF PIECES	DIAM. SCREW U	LENGTH L	REMARKS

TURN BUCKLES

MARK	NO. OF PIECES	DIAM. SCREW U	LENGTH X	LENGTH L	MARK

Fig. 9. Shop Bill for Sleeve Nuts and Turn Buckles

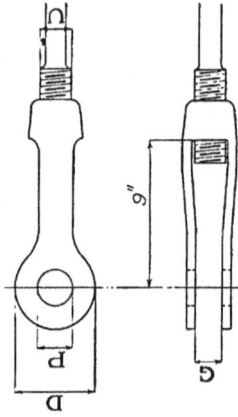

SALINE BRIDGE COMPANY

ORDER No. _____

SHEET No. __ - -

_____BRANCH

CLEVISES FOR

MARK	NO. OF PIECES	DIAM. D	SCREW U	PIN P	GRIP G	REMARKS

Fig. 10.   Shop Bill for Clevises

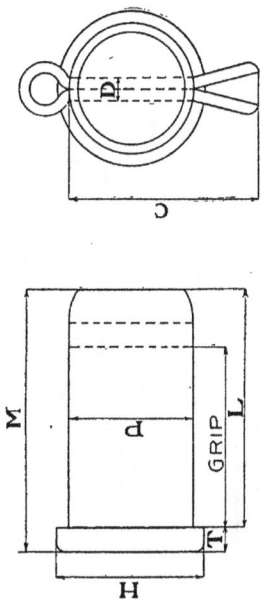

Fig. 11. Shop Bill for Cotters and Cotter Pins

SALINE BRIDGE COMPANY

ORDER No. - - - - - - - - - - -
- - - - - - - - - - -
- - - - - - - - - - - - - - - BRANCH

SHEET No. - - - -
- - - -
- - - -

6 Threads per inch

Screw D

Diam. S

Grip G

Length over all L

P

T

N

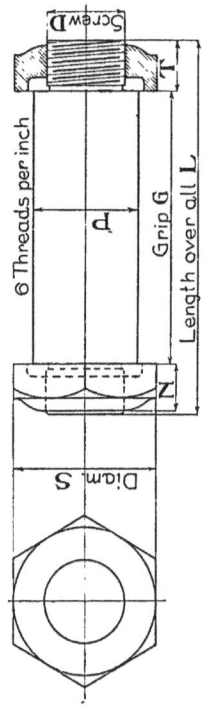

PINS AND LOMAS NUTS FOR

MARK	NO. PCS.	DIAM. P	GRIP G	SCREW		NUT		LENGTH OVER ALL L	NO. PILOT NUTS	NO. DRIVING NUTS	REMARKS
				DIAM. D	LENGTH T	DIAM. S	THICKNESS N				

Fig. 12. Shop Bill for Pins and Lomas Nuts

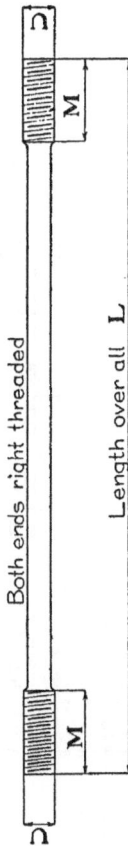

Fig. 13. Shop Bill for Upset Screw Rods

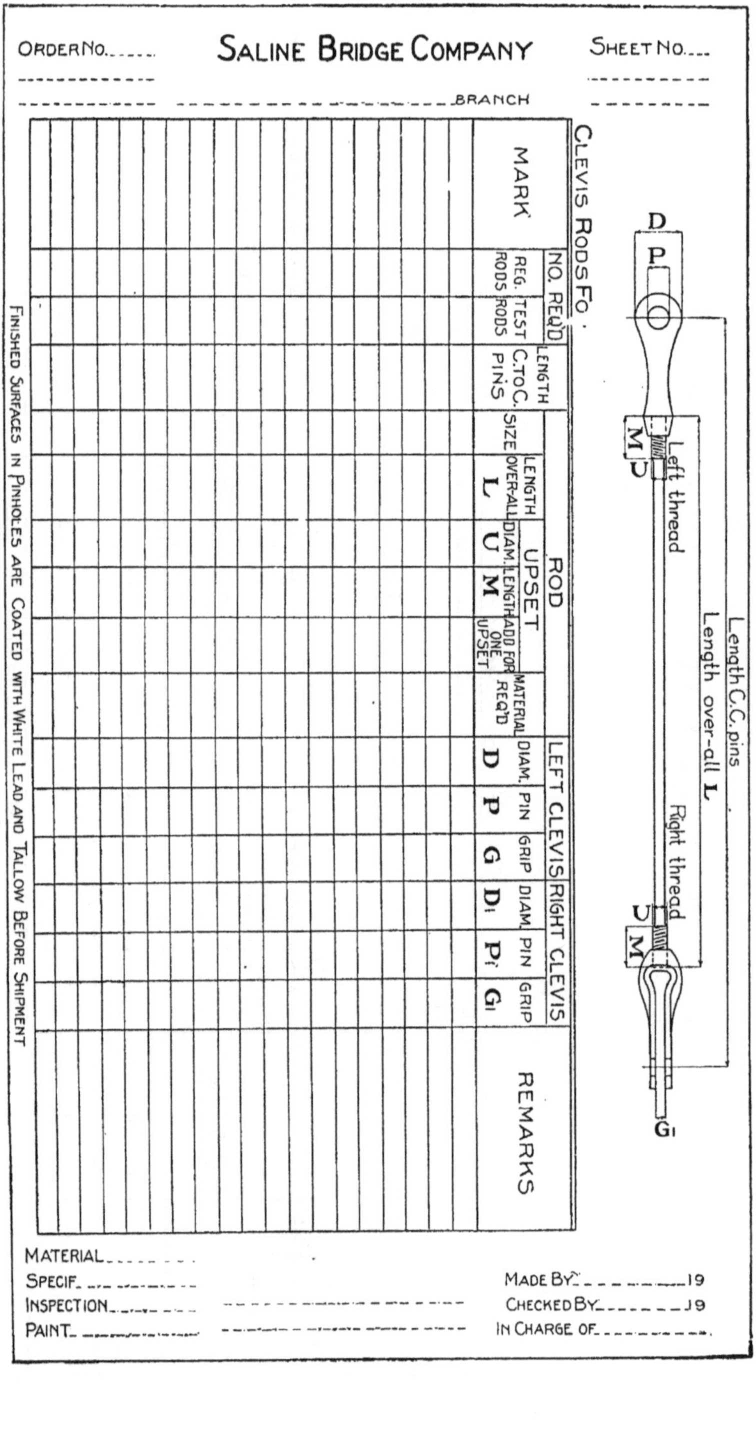

SALINE BRIDGE COMPANY

BRANCH

CLEVIS RODS FOR.

MARK	NO. REQ'D			LENGTH C. TO C. PINS	ROD				LEFT CLEVIS	RIGHT CLEVIS				REMARKS
	REG. RODS	TEST RODS	RODS PINS		LENGTH OVER ALL	SIZE DIAM. OVER ALL	UPSET	MATERIAL ADD'D FOR ONE UPSET REQ'D	DIAM. PIN	GRIP	DIAM. PIN	GRIP		
					L	U	M		D	P	G	D₁	P₁	G₁

FINISHED SURFACES IN PINHOLES ARE COATED WITH WHITE LEAD AND TALLOW BEFORE SHIPMENT

Fig. 14. Shop Bill for Clevis Rods

Length C.C. pins
Length over-all L
Left thread
Right thread

D P M U U M G₁

MATERIAL.........
SPECIF.........
INSPECTION.........
PAINT.........

MADE BY.........19
CHECKED BY.........19
IN CHARGE OF.........

SALINE BRIDGE COMPANY

BRANCH

Eye flattened where ordered

Length C to end **L**

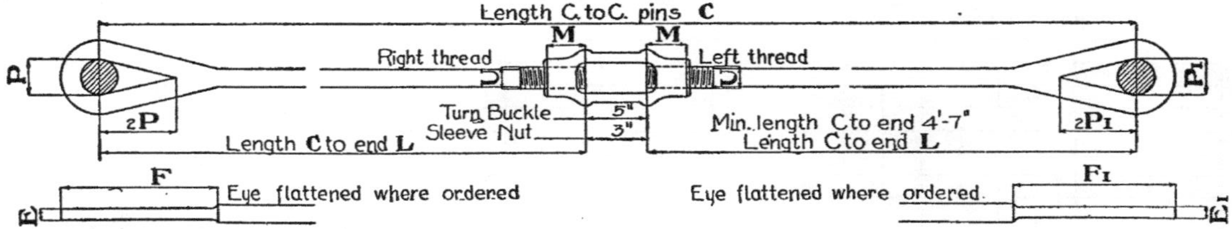

Right thread

Turn Buckle Sleeve Nut

Length C to C pins **C**

Left thread

Min. length C to end 4"-7"
Length C to end **L**

Eye flattened where ordered

LOOP RODS FOR					LONG END								SHORT END							
MARK	NO. REQ'D		SIZE OF RODS	REG. TEST OF ROD	LENGTH C TO C OF PINS	LOOP PIN FLATTEN			UPSET ADD			LENGTH C TO END REQ'D	MATER'L	LOOP PIN FLATTEN			UPSET			LENGTH C TO END REQ'D	MATER'L
	NO. RODS	RODS			C	P	E	F	U DIAM	M LENGTH	L			P	E₁	F₁	U DIAM	M LENGTH	L₁		

MATERIAL, WROUGHT IRON

SPECIF._____

INSPECTION_____

PAINT_____

MADE BY_____19____

CHECKED BY_____19____

IN CHARGE OF_____

FINISHED SURFACES IN PIN HOLES ARE COATED WITH WHITE LEAD AND TALLOW BEFORE SHIPMENT

Fig. 15. Shop Bill for Loop Rods

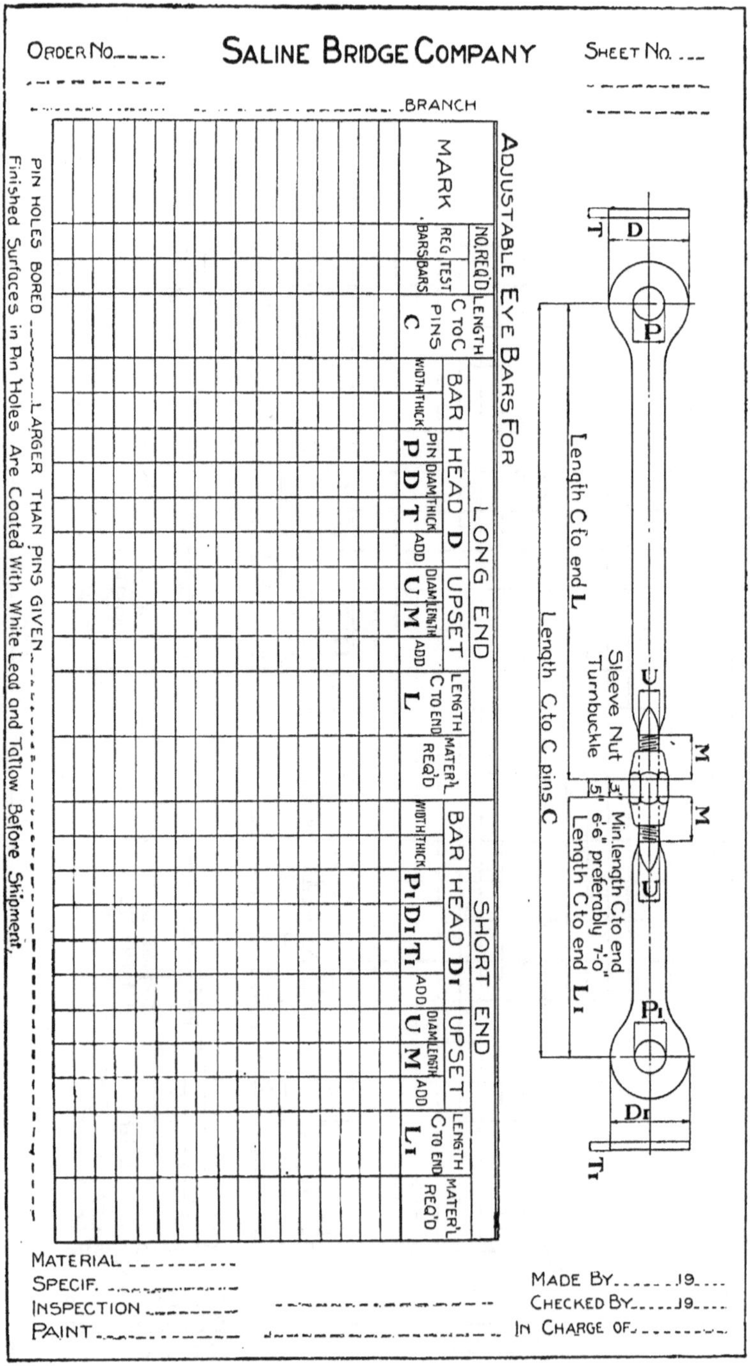

BRANCH

ADJUSTABLE EYE BARS FOR

LONG END

MARK	NO.REQ'D LENGTH		BAR HEAD D		UPSET			LENGTH MATER'L
	REG. BARS	TEST BARS	C TO C PINS C	WIDTH THICK P D	PIN DIAM. THICK T	DIAM. ADD U	LENGTH ADD M	C TO END REQ'D L

SHORT END

BAR HEAD D₁		UPSET			LENGTH MATER'L
WIDTH THICK P₁ D₁	PIN DIAM. THICK T₁	DIAM. ADD U	LENGTH ADD M	C TO END REQ'D L₁

PIN HOLES BORED _____ LARGER THAN PINS GIVEN _____
Finished Surfaces in Pin Holes Are Coated With White Lead and Tallow Before Shipment.

Fig. 16. Shop Bill for Adjustable Eye Bars

D
T
P

Length C to end **L**

Length C to C pins **C**

Sleeve Nut
Turnbuckle

3" Min. length C to end
6'-6" preferably 7'-0"
Length C to end **L₁**

M
M

U
U

P₁
D₁

T₁

MATERIAL _____
SPECIF. _____
INSPECTION _____
PAINT _____

MADE BY _____ 19 ____
CHECKED BY _____ 19 ____
IN CHARGE OF _____

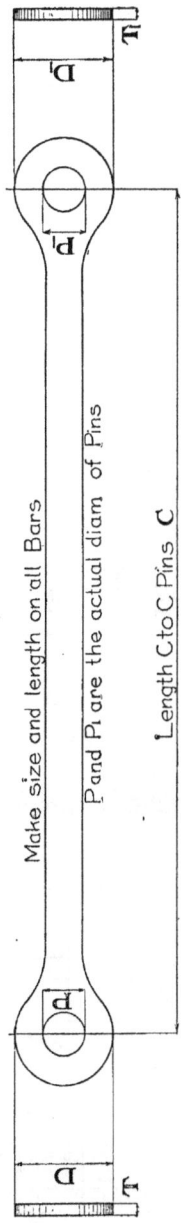

SALINE BRIDGE COMPANY

ORDER No. _____

SHEET No. _____

_____ BRANCH

Make size and length on all Bars

P and P1 are the actual diam. of Pins

Length C to C Pins C

(Labels on drawing: D, T, D1, T1, P, P1)

EYE BARS FOR

MARK	NO. REQ'D		SIZE		LENGTH C to C of PINS C	HEAD D				PIN P1	HEAD D1			MATERIAL REQUIRED
	REG. BARS	TEST BARS	WIDTH	THICK		PIN P	DIAM. D	THICK T	ADD		DIAM. D1	THICK T1	ADD	

Fig. 17. Shop Bill for Ordinary Eye Bars

SALINE BRIDGE COMPANY

ORDER No.--------

------------- BRANCH ------------

SHEET No.----------

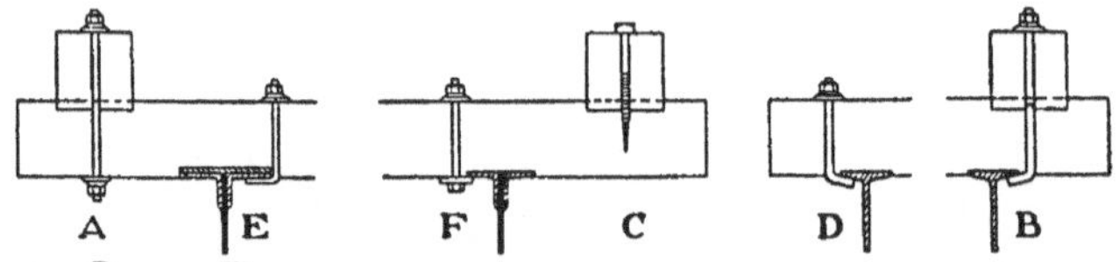

FLOOR BOLTS FOR

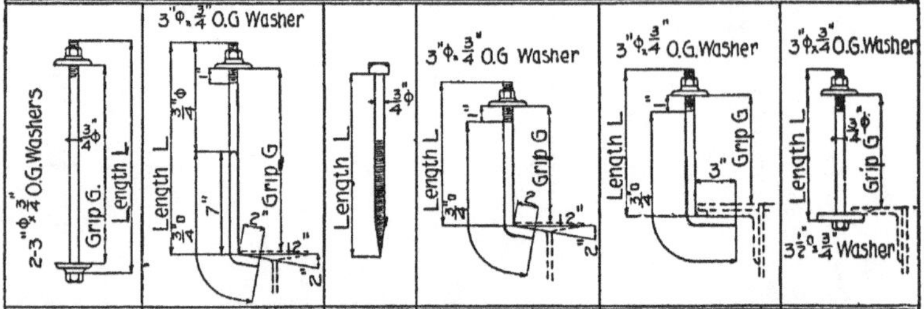

BOLT A			BOLT B			LAG SCREW C		BOLT D				BOLT E				BOLT F		
NO. REQ'D	GRIP G	LENGTH L	NO. REQ'D	GRIP G	LENGTH L	MATR'L REQ'D	NO. REQ'D	LENGTH L	NO. REQ'D	GRIP G	LENGTH L	MATR'L REQ'D	NO. REQ'D	GRIP G	LENGTH L	MATR'L REQ'D	NO. REQ'D	GRIP G	LENGTH L

Fig. 18. Shop Bill for Floor Bolts

SALINE BRIDGE COMPANY

ORDER No. -------

------------- BRANCH ------------

SHEET No. --------

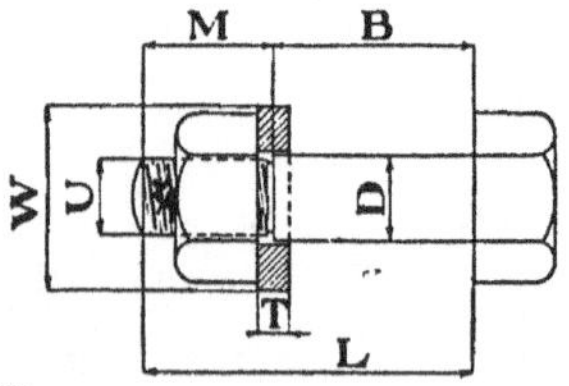

TURNED BOLTS FOR

MARK	NO. OF PCS.	DIAM. OF BOLT D	SCREW			LENGTH OF BOLT L	WASHER		REMARKS
			LENGTH B	LENGTH M	DIAM. U		DIAM. W	THICK T	

Fig. 19. Shop Bill for Turned Bolts

SALINE BRIDGE COMPANY

ORDER No.

SHEET No.

BRANCH

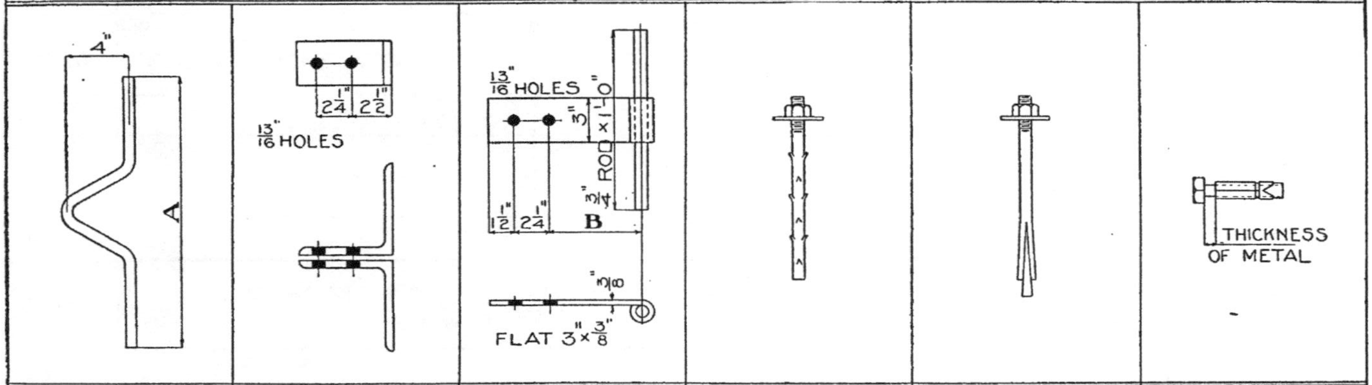

THICKNESS OF METAL

FLAT 3×8

$\frac{13}{16}$" HOLES

$1\frac{1}{24}$" $1\frac{1}{22}$"

$1\frac{1}{24}$" $2\frac{1}{24}$"

FLAT

$1\frac{1}{24}$" $1\frac{1}{22}$"

$\frac{13}{16}$" HOLES

4"

A

B

ANCHOR NO.1			ANCHOR NO.2			ANCHOR NO.3		HACKED BOLT				SPLIT BOLT				EXPANSION BOLT		
NO. OF PCS.	DIAM.	DIST. A	LENGTH OF MAT'L	NO. OF ANG'S	SIZE OF ANGLES	LENGTH OF ANGLES	NO. OF ANCH.	DIST. B	LENGTH OF FLAT	NO. OF PCS.	DIAM.	LENGTH	SIZE OF WASHER	NO. OF PCS.	DIAM.	LENGTH	SIZE OF WASHER	NO. OF PCS.	DIAM.	LENGTH	THICK'S OF METAL

Fig. 20. Shop Bill for Anchors and Anchor Bolts

SALINE BRIDGE COMPANY
SHOP BILL

ORDER ASSIGNED TO __Columbus_____PLANT

NAME OF STRUCTURE __Bridge over Branch of Ten Mile Creek__

NAME OF CUSTOMER __Nelson and Buchanan Co. (ord. 427.)__

WORK FABRICATED AT

__Columbus_____PLANT

NO. OF PIECES	DESCRIPTION	LENGTH FEET	LENGTH INCHES	PIECE MARK	CALC. WT. ONE PIECE / TOTAL	REMARKS	NO. OF PIECES	SECTION	WT PER FT IN LBS.	LENGTH FEET	LENGTH INCHES	ITEM NUMBER
4	Floor Beams				F							
4	I-beams	15	$9\frac{1}{4}$					15"	42	15	$9\frac{1}{4}$	
32	Lug Ls	0	$7\frac{1}{2}$					$3 \times 2\frac{1}{2} \times \frac{1}{2}$				
16	"	0	5					"				
16	End Conn. Ls	0	10					$4 \times 3 \times \frac{3}{8}$				
	2 End Struts				S							
2	Channels	18	0					6"	8	18	0	
56	Bearing Pls.	0	5					$3 \times \frac{1}{2}$				

Fig. 21. Shop Bill for Members of Bridge

SALINE BRIDGE COMPANY

Order Assigned to ....Columbus....... PLANT

Work Fabricated at ....Columbus.... PLANT

Name of Structure.....Neff Bridge on Cline Free Pike...........

Name of Purchaser....Huston and Cleveland.........

Erector's List of Field Rivets and Bolts.

NUMBER REQUIRED	DISCRIPTION	DIAM.	GRIP	LENGTH	SHAPE OF HEAD	PIECES CONNECTED AND REMARKS
52	Bolts	$\frac{5}{8}$	$\frac{5}{8}$	$1\frac{1}{2}$	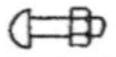	Buckle Pls and P2 to Girders GIR & L
8	"	"	$\frac{15}{16}$	$1\frac{3}{4}$	"	" " " " " " "
192	"	"	$\frac{3}{4}$	$1\frac{1}{2}$	. "	" " " " Stringers S1 & S2
64	"	"	$1\frac{1}{16}$	2	"	" " " " " " "
50	"	"	$\frac{5}{8}$	$1\frac{1}{2}$	"	" " to PLs P2
2	"	$1\frac{1}{2}$	$\frac{7}{8}$	$1\frac{1}{2}$	- "	Name Plate to GIR
4	Rivets	$\frac{5}{8}$	$\frac{11}{16}$	$2\frac{1}{2}$		B2 to S2
2	"	"	$\frac{1}{2}$	$1\frac{7}{8}$	"	B2 to GIR & L
4	"	"	$\frac{1}{2}$	$1\frac{7}{8}$	"	B2 to BL
8	"	"	$\frac{1}{2}$	$1\frac{7}{8}$	"	B1 to Railing on GIR & L

Fig. 22. Erector's List of Field Rivets and Bolts

ORDER No._____

_____ BRANCH

SHEET No._____

SALINE BRIDGE COMPANY

RIVETS AND BOLTS FOR

RIVETS (PLAIN)				RIVETS (COUNTERSUNK)				BOLTS				
NUMBER REQUIRED	DIAM.	LENGTH UNDER HEAD	REMARKS	NUMBER REQUIRED	DIAM.	LENGTH OVER ALL	REMARKS	NUMBER REQUIRED	DIAM.	LENGTH UNDER HEAD	SHAPE OF HEAD	REMARKS

Fig. 23. Shop Bill for Field Rivets and Bolts

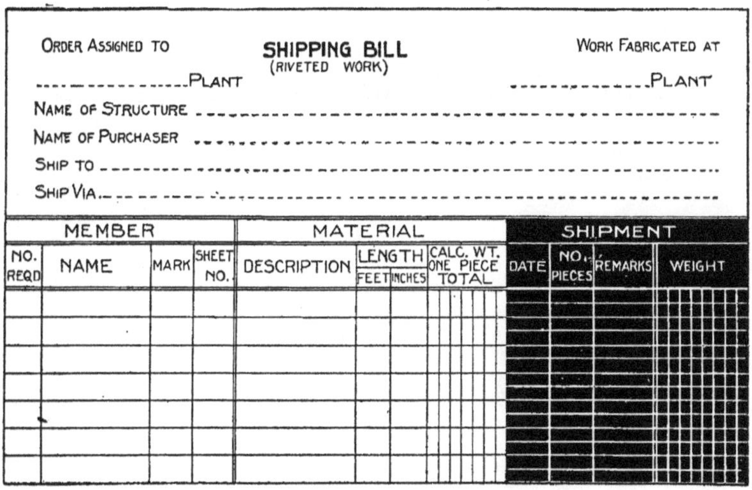
ORDER ASSIGNED TO

SHIPPING BILL
(RIVETED WORK)

WORK FABRICATED AT

.............................PLANT ....PLANT

NAME OF STRUCTURE ....

NAME OF PURCHASER ....

SHIP TO ....

SHIP VIA ....

MEMBER				MATERIAL				SHIPMENT			
NO. REQD	NAME	MARK	SHEET NO.	DESCRIPTION	LENGTH FEET	LENGTH INCHES	CALC. WT. ONE PIECE TOTAL	DATE	NO. PIECES	REMARKS	WEIGHT

Fig. 24. Shipping Bill

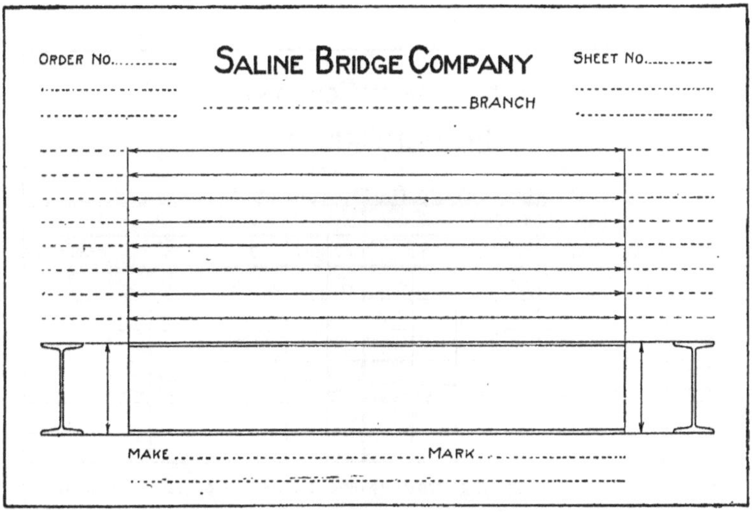
ORDER NO...........

SALINE BRIDGE COMPANY

SHEET NO...........

..............................BRANCH

MAKE ....MARK..............................

Fig. 25. Shop Bill for I-Beams

ORDER ASSIGNED TO   **SALINE BRIDGE COMPANY**   WORK FABRICATED AT

................................PLANT   ....PLANT

NAME OF STRUCTURE..

NAME OF PURCHASER..

Fig. 26. Shop Bill with Blank Space for Sketch

SALINE BRIDGE COMPANY
CHANGE ORDER

PLEASE MAKE THE FOLLOWING CHANGES

			CHANGE FROM							TO				
ITEM	NO	KIND	SIZE	LENGTH FT. IN.	MARK	SKETCH NO	NO	KIND	SIZE	LENGTH FT. IN.	MARK	SKETCH NO
1013	10	I	8"x20.5#	17 10½			6	I	8"x20.5#	17 10½		

FINISHED SURFACES IN PIN HOLES ARE COATED WITH WHITE LEAD AND TALLOW BEFORE SHIPMENT.

MATERIAL..............

SPECIF................

INSPECTION........... ....

PAINT............. ....

MADE BY.........19...

CHECKED BY.......19...

IN CHARGE OF...........

Fig. 27. Sheet Used for Change of Order

DETAILING — GENERAL INSTRUCTIONS

Lettering. In order that the drawing may give the necessary information and that no mistakes should occur in the reading of the drawing by the shopmen or others, it is necessary that the letters and dimensions upon the drawing be made so that they are exceed-

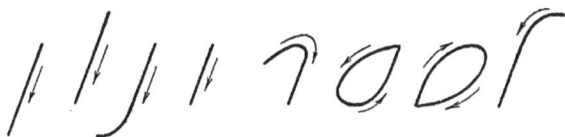

Fig. 28.   Method of Constructing Parts of Small Arc-Line Letters

ingly clear. In order to save time in lettering, an alphabet should be used that can be made quickly and easily. The alphabet which is known as the straight-line alphabet fulfills these conditions. It

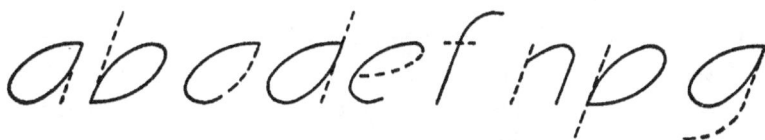

Fig. 29.   Method of Constructing Parts of Small Arc-Line Letters

is made by one of the characters or by a combination of the characters shown in Fig. 28.   A study of Fig. 29 will show that the general scheme of this system consists of the oval and the straight line.

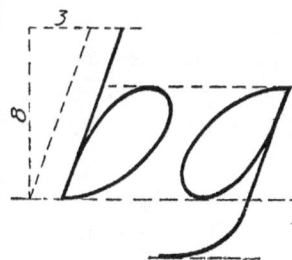

Fig. 30.   Method of Constructing Parts of Small Arc-Line Letters

The slant at which these letters are made is a very important factor in a drawing, the proper slant being 3 in 8, as shown in Fig. 30.   Even a slight increase, however, will give one the impression that the letters lean too far forward and it will spoil the appearance of a drawing otherwise good.

The height of the lower part of the letter should be equal to two-thirds or more of the total height.   Figures should be of the same height as the capital letters.   The total height of the small letters should not be less than one-tenth of an inch.   This makes the capitals three-twentieths of an inch high, not less. The reason for adopting this height of letters is in order that, if necessary, ordinary tracings may be reduced for publication and the letters will then show up clearly. Fig. 31 shows the complete alphabet and

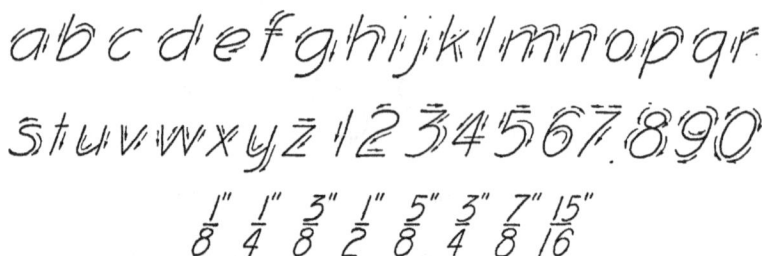

Fig. 31. The Completed Letter, with Arrows Showing Direction of Stroke

the numerals from 1 to 0, also several fractions. The fractions should never be made less than one-tenth of an inch in height for each of the members, and the dividing line should be horizontal, never slanting. Fig. 31 also shows by means of small arrows the direction the stroke should take when making the different letters and figures.

There is a tendency to make several of the letters and figures as shown in Fig. 32. This tendency should be carefully avoided, special attention being called to the turned-up ends of the members of different characters. Care should be taken not to get the upper part of the s and the 8 larger than the lower part. If this is done or if the two parts are made equal the upper will appear to be much larger and these characters will look out of proportion.

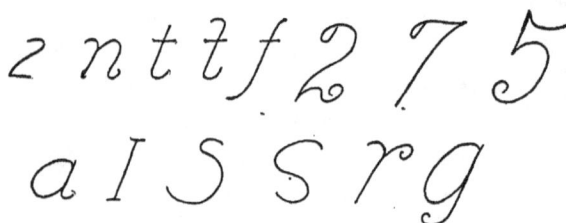

Fig. 32. Example of Poorly Constructed Letters

The capital letters S, G, E, F, P, and R, and the figures 2 and 5, present some difficulties. These characters are shown in Fig. 33, and may briefly be commented on as follows:

Letter S. The letter S should begin at the point 1 slightly inside of the circumscribed parallelogram. The line should then be tangent to the top and should come slightly inside of the further side at point 2. It should then cross the center line above the middle height at the point 3 and be tangent at point 4 and point 5 as indicated.

. *Letter G.* The letter G should start at the right side of the parallelogram and be tangent to the top, left side, and bottom as well as the right-hand side where it extends upward to a height of one-half of its total height before the horizontal line, which should extend one-half of the distance across the letter, is drawn.

Letter E. The letter E presents no difficulties other than a central horizontal line should extend about two-thirds of the distance

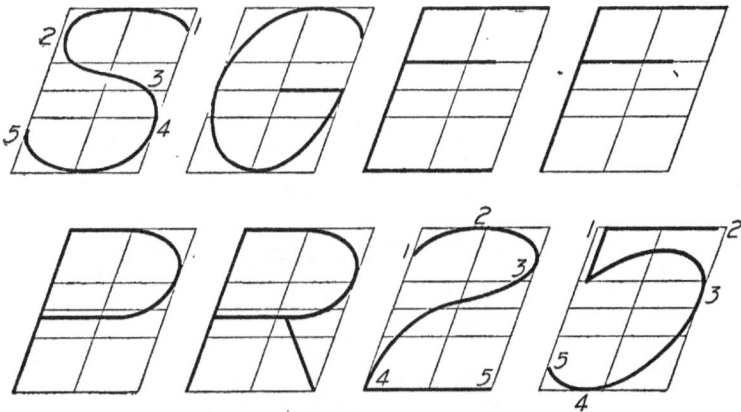

Fig. 33. Proportion and Slant of Capitals

across the letter and should be at an elevation of two-thirds the height.

Letter F. The letter F is but a part of the letter E as indicated.

Letters P and R. P and R are constructed on the same general principle. The upper part of both letters should be at least one-half or more of the total height, and in the case of R the lower right-hand stroke should not extend further than the right-hand side of the circumscribed parallelogram.

Figure 2. The figure 2 is constructed by starting at the left-hand side of the circumscribed parallelogram and continuing tangent as indicated in the figure at points 2, 3, and 4. The lower part 4—5 should in all cases be horizontal and it should never extend further than the right-hand side of the circumscribed parallelogram.

Figure 5. The figure *5* should start at the point 1 and extend downwards one-third of the total height. The lower part of the figure should then be drawn, being tangent at point 3 and 4 and slightly curled up at 5 where it should extend a little further to the left of the upper part. The horizontal part 1—2 should not extend quite up to the right-hand side of the circumscribed parallelogram.

In all cases where inch or foot sizes are employed, they should be made clearly and regularly and should be not less than one-twentieth of an inch in length.

Letters and figures should always be made by beginners by first preparing guide lines drawn with a pencil. Even in case the

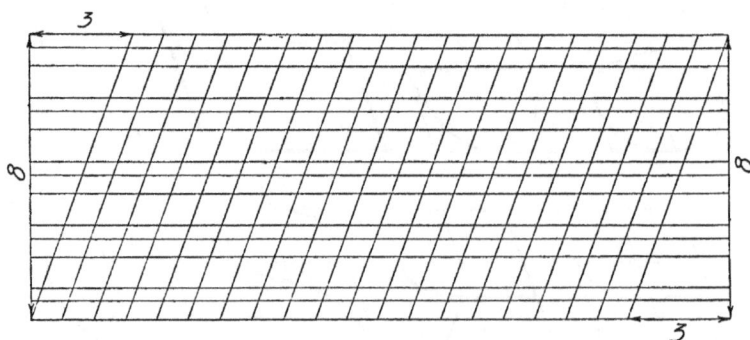

Fig. 34.   Guide Sheet for Obtaining Correct Slant in Letters

guide lines have been drawn upon the detailed paper, it is also advisable to draw them upon the tracing cloth, or place under the cloth a sheet similar to Fig. 34, with the lines drawn in ink, to act as a guide. This practice should be continued until enough skill has been acquired to make the letters uniform, without the assistance of more than a line or two.

The only manner in which a person can become proficient in lettering is through practice. A piece of paper ruled up and having the slant of the letter placed upon it as shown in Fig. 34 will be found an excellent thing on which to practice lettering. Letters can not be made nicely and quickly as one would suppose. Care and time are required until the draftsman becomes proficient in this respect.

TABLE IV
Abbreviations

SYMBOL	SIGNIFICANCE
L. or Ls.	Angle or angles
[. or [s.	Channel or channels
I. or Is.	I beam or beams
Z. or Zs.	Zee bar or bars
T. or Ts.	Tee beam or beams
Pl., Plt., or Pls., Plts.	Plate or Plates
@	at
Fill.	Filler
Stiff.	Stiffeners
Fl. or Flg.	Flange
r.	Rivet
f. r.	Field rivets
s.r.	Shop rivets
e.	Eccentricity
c.l. or ₵	Center line
o or φ	Diameter
#	Pound or pounds
c. to c. or ₵'s	Center to center
Latt. or Latts.	Latticed or lattices
Lat. or Lats.	Lateral or laterals
alt.	Alternate
M. Pl.	Masonry plate
Spl.	Splice

Abbreviations. In the making of drawings certain abbreviations are used in order to save time and for the sake of convenience in many other respects. These abbreviations together with what they signify are given in Table IV. They should be carefully studied and should be written close to the material to which they apply and should at least be one-sixth to one-eighth of an inch from the material. Never write dimensions or letters so close to a line that they will interfere with the line. In writing dimensions at a considerable distance from the piece of material or place to which they apply, an arrow is used to indicate their proper position. In all such cases the arrow head should be at the end of the line which points to the place to which the abbreviation or dimension applies.

Fig. 35 illustrates some cases and also shows the form which the arrow should take in order to present a good appearance on the drawing.

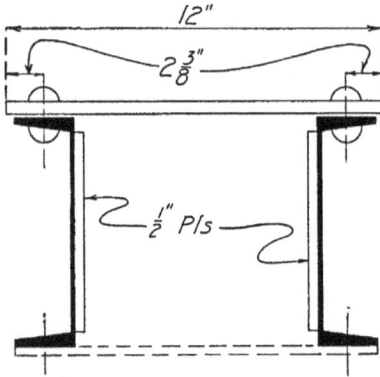

Fig. 35.　Correct Use of Arrow and Line in Dimensioning

Dimension and Material Notation. *Proper Placing.* A drawing may be said to have been correctly dimensioned when any desired necessary dimensions may be obtained from it without it being required that any dimensions should be added or subtracted or divided in order to obtain the desired result, and when no unnecessary dimensions are upon the drawing. By necessary dimensions are meant those dimensions which are required in order that the material may be fabricated so that the finished struc-

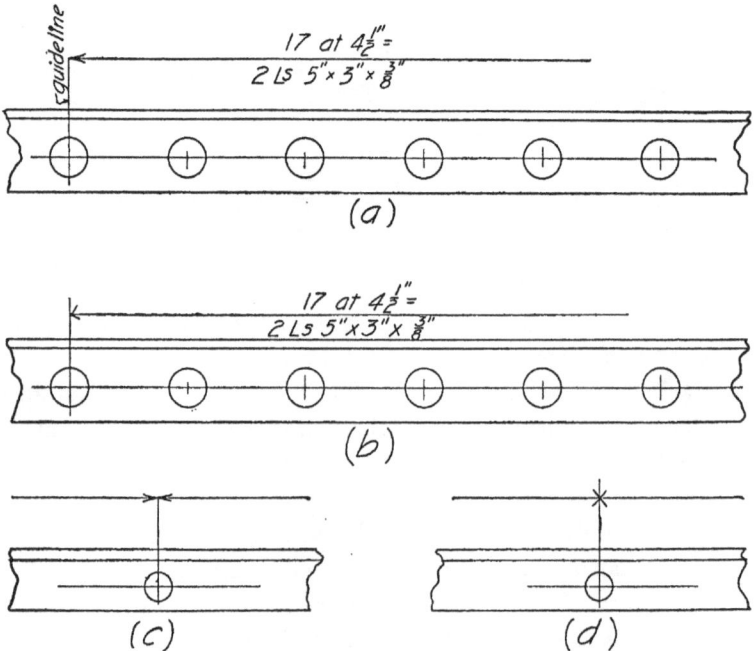

Fig. 36.　Examples of Arrow-Head Construction and Proper Location of Dimensions

ture is as desired. Dimension lines should be full, not dotted or dashed; guide lines, which are lines indicating the limits of the dimensions, should not extend beyond the dimension line. The dimensions should be placed where possible above the line and should not, as mentioned before,

Fig. 37. Correct Arrow Head

touch the line at any place. Dimension lines should be far enough from the piece which they dimension in order that the letters and figures indicating the character of the material and its size may be placed between the dimension line and the material itself. Fig. 36a shows good practice and Fig. 36b poor practice.

Arrow heads are a source of trouble and should be made with care if the drawing is to present a good appearance when finished. They should be made as indicated in Fig. 36a and Fig. 36c, and not as in Figs. 36b and 36d. They should not consist as indicated in the

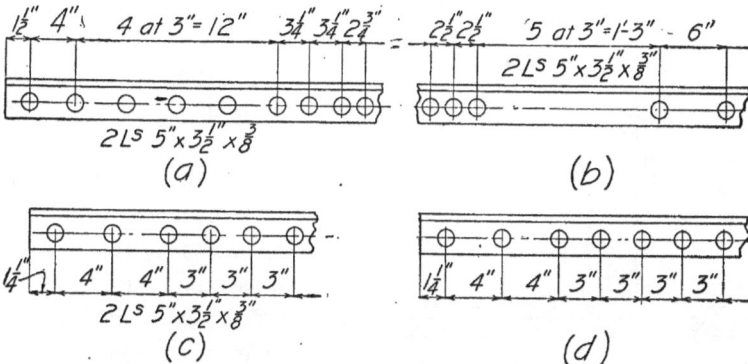

Fig. 38. Correct and Incorrect Placing of Dimensions

figure showing the wrong construction, of a cross or half cross of a straight or nearly straight line, but should have a gradual slope as in Fig. 37 where it is greatly exaggerated.

Dimensions as mentioned above should be placed above the dimension line where possible and the material should be noted so as not to interfere with the dimensioning. Figs. 38a and 38b show good practice and Figs. 38c and 38d poor practice. Sometimes it is necessary to place the dimensions as in Fig. 38c and 38d but never place the material notation as shown in the same figures. Fig. 38b gives the preferable method.

When several spaces are equal, the matter may be written as
so many spaces at so much is equal to so much, or each space may
be dimensioned separately as shown in Figs. 38a and 38b. In case
the space is too small to write a single dimension in it clearly, the
dimension may be put at one side and an arrow used to show where
it belongs, Fig. 35.

In writing dimensions the inches should be given as well as the
feet, and in case the inches amount to nothing or to a fraction, a
cipher should take the place of the inches.

It is not necessary when all rivets are shop rivets to draw in each
in such cases to put in the end rivets and to inidicate the spacing
and every rivet when the spacing is the same. It is only necessary of
those which lie between but which are not shown. Fig. 38b illus-

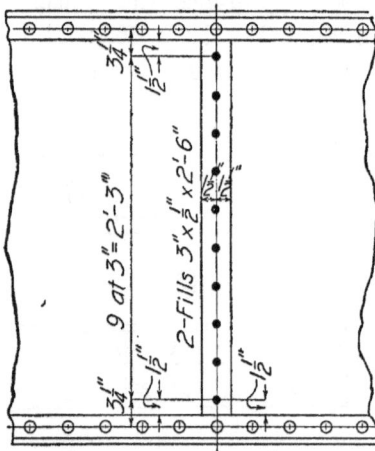

Fig. 39. Correct Method of Indicating Shop Rivets

trates this. In case of field rivets
all rivets must be shown. No de-
parture from this rule should be
allowed. Fig. 39 is an example of
this. It is noted in this figure that
although the spacing of many of
the rivets is the same, yet all are
shown in their proper place.

In placing dimensions where
two or more members are detailed
together, dimensions for the main
member should run straight
through from one end to the
other. The dimensions of the
larger member in so far as they
are the same as the dimensions for the smaller may be used for the
smaller member and additional or subdimensions be placed in convenient places in order to complete the detailing of the smaller member. As an example of this see Fig. 38a where the edge distances
and the method of detailing should be noted, and Fig. 39 also where
the edge distances are the subdimensions. In Fig. 39 two dimensions are given at one of the ends. This illustrates two methods
of placing the same dimension. The dimension directly under the
line of dimensions for the main member is placed in the preferable
way. In the placing of subdimensions great care should be taken

not to make them too small or to place them so that they interfere with the guide lines of the main dimension.

Notation Used. As stated before, the feet and inches should always be given when a multiple space is given. For example they should be written thus:

$$5 \ @ \ 6'' = 3'\text{-}0''$$
$$2 \ @ \ 3'' = 0'\text{-}6''$$
$$7 \ @ \ 4'' = 2'\text{-}4''$$

In single dimensions less than 1 foot it is not necessary to state the 0 for the foot, and, therefore, we have for example 4″, 6″, 11″, etc., up to and including 12″. A dash should always be placed between the feet and inches as shown above. Careful attention should be paid to this detail since the omission of the dash may cause the dimensions to be considered as all feet or all inches and time and money will accordingly be lost.

In material notation the following are the rules:

For angles the number should be placed first, the angle sign second, the dimension of the greatest leg next, then the small leg, then the thickness, and then the length in feet and inches.

For example,

$$2\text{-}Ls \ 5'' \times 3'' \times \tfrac{3}{8}'' \times 17'\text{-}2''$$

For plates the number comes first, the abbreviation next, the width in inches next, then the thickness in inches, and finally the length in feet and inches.

For example,

$$1\text{-}Pl. \ 18'' \times \tfrac{3}{8}'' \times 2'\text{-}4''$$

For beams and channels the number is stated first, next the depth in inches, then the weight in pounds per foot, then the sign, and finally the length in feet and inches.

For example,

$$2\text{-}12'' \times 31\tfrac{1}{2}{}^{\#} \ Is \times 18'\text{-}2''$$
$$5\text{-}7'' \times 9\tfrac{3}{4}{}^{\#} \ Cs \times 16'\text{-}5\tfrac{1}{4}''$$

Zee bars are designated by their depth and thickness. The number is written first, the depth next, then the thickness, then the sign and finally the length in feet and inches.

TABLE V

Dimensions and Conventional Representation of Rivets

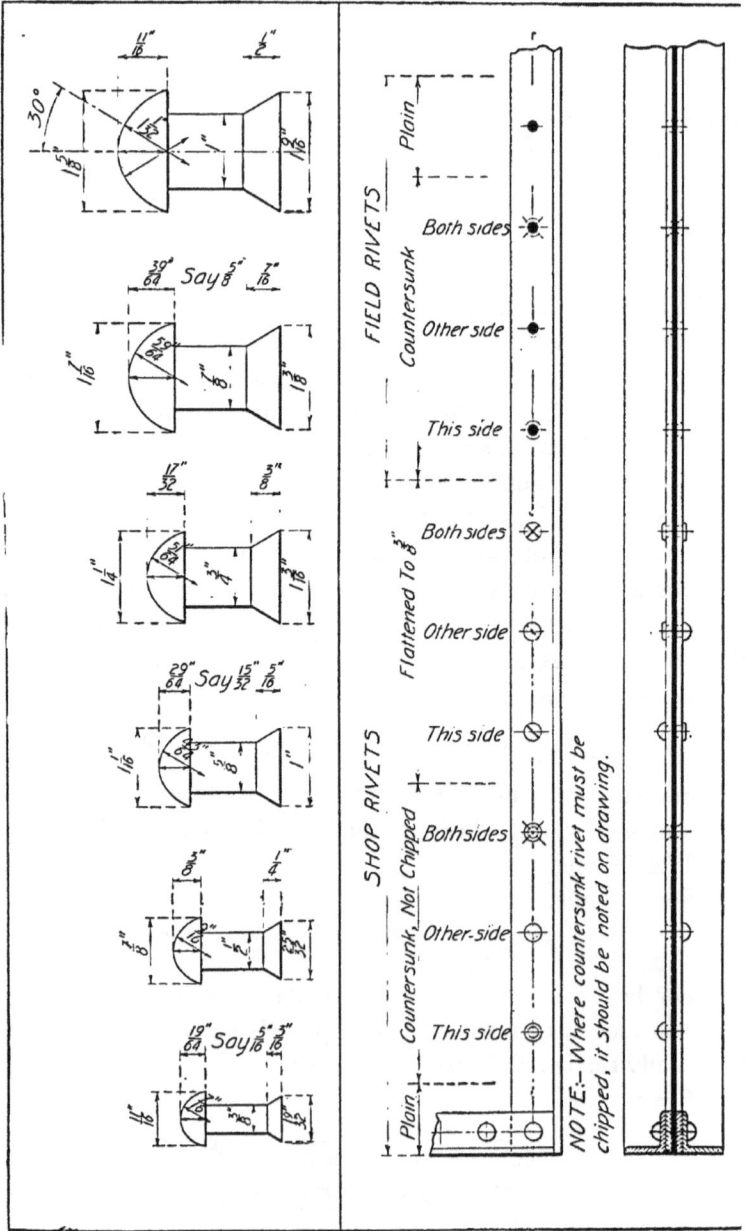

SHOP RIVETS

| This side | Other-side | Both sides | | Plain |
| Countersunk, Not Chipped | | | | |

FIELD RIVETS

| This side | Other side | Both sides | | Plain |
| Flattened to $\frac{5}{8}''$ | | Countersunk | | |

NOTE:- Where countersunk rivet must be chipped, it should be noted on drawing.

For example,

$$3 - 6'' \times \tfrac{3}{8}'' \ Zs \ \times 14' - 8''$$

Bars are designated by their number, then their size, or diameter, and finally their length in inches.

For example,

$$3 - 1'' \square \ \times \ 20' - 2''$$
$$1 - \tfrac{3}{4}'' \phi \ \times \ 16' - 4\tfrac{1}{4}''$$

Rivets and Rivet Spacing. Rivets are made in various sizes and are spoken of according to the diameter of their shank. Thus a $\frac{7}{8}$-inch rivet is one which has its shank $\frac{7}{8}$ inch in diameter. The heads of the rivets are not perfect hemispheres, being less in height than one half the diameter of the head. Table V gives the dimensions of rivets of various diameters and their conventional representation in detail drawings. These dimensions are desirable on the drawings since they are often necessary in order to so figure the work that the material will not strike the heads. Rivets smaller than $\frac{3}{4}$ inch are seldom used except where the size of the material requires it. Rivets larger than $\frac{7}{8}$ inch in diameter are seldom used except in the heaviest work; and the beginner is advised not to use them until he has permission from those above him in charge.

Rivets should not be placed so close that the material between them is unduly injured by pushing or that the driving tool or "dolly" will interfere with one rivet when driving the other; likewise they should not be placed so far apart that the material between them will separate or open up. Unless specified otherwise in the specifications Table VI may be taken as good practice; for $\frac{3}{4}$-inch, $\frac{7}{8}$-inch, and 1-inch rivets, the minimum spacing is seen to be three diameters of the rivet.

TABLE VI
Minimum Rivet Spacing
(All dimensions given in inches)

SIZE of RIVETS	$\frac{1}{4}$	$\frac{3}{8}$	$\frac{1}{2}$	$\frac{5}{8}$	$\frac{3}{4}$	$\frac{7}{8}$	1
Minimum Spacing Center to Center	1	$1\frac{1}{4}$	$1\frac{3}{4}$	2	$2\frac{1}{4}$	$2\frac{5}{8}$	3

The maximum spacing allowable is usually sixteen times the thickness of the thinnest plate they go through. The minimum and maximum limits placed above are not to be used wherever possible. Few engineers consider it advisable or permit spacings less than $2\frac{1}{2}$ inches and 3 inches, or more than 4 inches and 5 inches for $\frac{3}{4}$-inch and $\frac{7}{8}$-inch rivets, respectively.

The minimum limits above refer to the center to center of rivets, while the maximum values refer to the distance center to

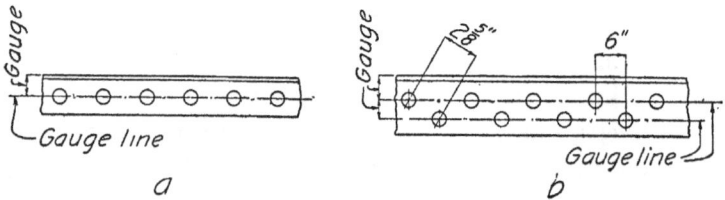

Fig. 40. Angles with Gauge Line

center measured along the gauge line or line along which the rivets are placed.

Gauge lines may be single, Fig. 40a, or double as in Fig. 40b. The gauge of a shape is the distance of the gauge line from a certain base. In the angle it is the back, in the channel it is the back, while in the I-beam it is the bisecting line of the web.

The gauges for standard channels and I-beams are given in the handbooks of manufacturers, such as Cambria, Carnegie, etc.,

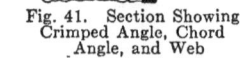

Fig. 41. Section Showing Crimped Angle, Chord Angle, and Web

which books also give the size of rivet or bolt which can be used in the flange of any certain I-beam or channel. This does not mean that the size of bolt or rivet there given must be used in the web also, in fact, $\frac{3}{4}$-inch and $\frac{7}{8}$-inch should be used in the web, no matter what size is specified for the flange.

The standard gauges for angles are given in Table VII.

While a double gauge is shown for a 5-inch leg, it is very undesirable to use it. Do not use 5-inch legs with double gauge lines. Likewise, do not use a single-gauge line on an angle with a 6-inch leg or more, unless specially told to do so by those higher in authority.

TABLE VII
Standard Gauges for Angles
(All dimensions given in inches)

L	g	Maximum Rivet or Bolt	L	g	Maximum Rivet or Bolt	L	g	Maximum Rivet or Bolt
8	$4\frac{1}{2}$	$\frac{7}{8}$	$3\frac{1}{2}$	2	$\frac{7}{8}$	2	$1\frac{1}{8}$	$\frac{1}{2}$
7	4	$\frac{7}{8}$	3	$1\frac{3}{4}$	$\frac{7}{8}$	$1\frac{3}{4}$	1	$\frac{1}{2}$
6	$3\frac{1}{2}$	$\frac{7}{8}$	$2\frac{3}{4}$	$1\frac{5}{8}$	$\frac{3}{4}$	$1\frac{1}{2}$	$\frac{7}{8}$	$\frac{3}{8}$
5	3	$\frac{7}{8}$	$2\frac{1}{2}$	$1\frac{3}{8}$	$\frac{5}{8}$	$1\frac{1}{4}$	$\frac{3}{4}$	$\frac{3}{8}$
4	$2\frac{1}{4}$	$\frac{7}{8}$	$2\frac{1}{4}$	$1\frac{1}{4}$	$\frac{5}{8}$	1	$\frac{9}{16}$	$\frac{1}{4}$

L	g_1	g_2	L	g_1	g_2
8	3	3	6*	$2\frac{1}{2}$	$2\frac{1}{4}$
7	$2\frac{1}{2}$	3	5	2	$1\frac{3}{4}$
6	$2\frac{1}{4}$	$2\frac{1}{2}$			

*When thickness is $\frac{3}{4}$ inch or over.

In the spacing of rivets in crimped angles, the distance "b", Fig. 41, should be $1\frac{1}{2}$ inches plus twice the thickness of the chord angles, but never less than 2 inches.

The grip of a rivet is the length under heads after the rivet has been driven. The length of a rivet is the length of the shank before the rivet is driven, Fig. 42, these lengths for various grips being easily found in any manufacturer's handbook.

Care should be taken in case of castings to add $\frac{1}{8}$ inch more to those values given.

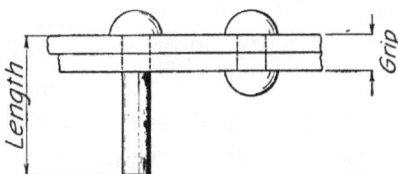

Fig. 42. Rivet Before and After Driving

Rivets may have two full heads or may have one or both heads countersunk or flattened or any combination. Such conditions are signified by certain signs, all of those in common use being listed in the handbooks already referred to, and also shown in Table V.

A rivet can be driven as close to a projection as one-half the diameter of the head plus $\frac{1}{4}$ inch. This requires a special "dolly". The dolly generally used requires $\frac{2}{3}D+\frac{1}{4}$ inch. This is about $1\frac{1}{4}$ inches for a $\frac{7}{8}$-inch rivet and about $1\frac{1}{8}$ inches for a $\frac{3}{4}$-inch; see Fig. 43 and Table V. In some instances a special gauge, that is, one other than given in Table VII, is used. In such cases care should be taken to see that the distance A, to the fillet, or curve of the angle, is sufficient, otherwise the dolly could not come down evenly and an imperfect head is the result.

When rivets are staggered, it is necessary to know how close they may be spaced in order that they may not be less than the minimum allowed distance center to center. Table VIII gives the distances center to center of rivets for given values of the spacing and gauge line. The distances below and to the right of the upper zigzag line are large enough for $\frac{3}{4}$-inch rivets while those below and to the right of the lower zigzag line are large enough for $\frac{7}{8}$-inch rivets. For example, if the gauge "g" was $1\frac{3}{4}$ inches, the spacing must be at least 2 inches in order that the distance center to center would not be less

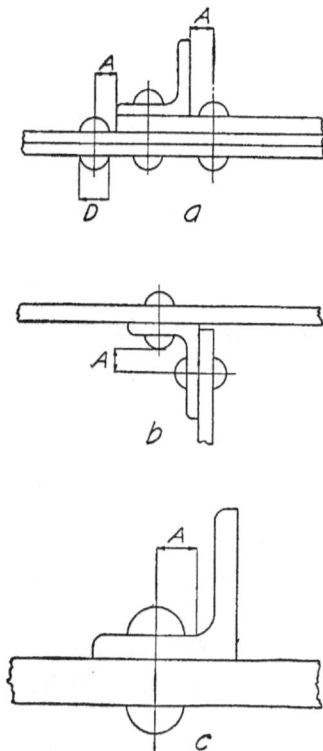

Fig. 43.   Diagram for Minimum Rivet Spacing

than $2\frac{5}{8}$ inches, the rivets being $\frac{7}{8}$ inch. If the rivets were $\frac{3}{4}$ inch, the spacing must be $1\frac{1}{2}$ inches or more in order to have the distance center to center not less than $2\frac{1}{4}$ inches. These values are found by going down from the value $1\frac{3}{4}$ inches in the top row until a value equal to or just greater than the $2\frac{5}{8}$ or $2\frac{1}{4}$ inches is found, and then following across to the first column where required spacing is found.

TABLE VIII

Values Center to Center for Various Spacings

(All dimensions in inches)

VALUES OF s (inches)	VALUES OF x FOR VARYING VALUES OF g AND s — VALUES OF g (inches)												
	7/8	1	1 1/8	1 1/4	1 3/8	1 1/2	1 5/8	1 3/4	1 7/8	2	2 1/8	2 1/4	2 3/8	2 1/2
1 1/8	1 7/16	1 1/2	1 9/16	1 11/16	1 3/4	1 7/8	2	2 1/16	2 3/16	2 5/16	2 3/8	2 1/2	2 5/8	2 3/4
1 1/4	1 1/2	1 5/8	1 11/16	1 3/4	1 7/8	1 15/16	2 1/16	2 1/8	2 1/4	2 3/8	2 7/16	2 9/16	2 11/16	2 13/16
1 3/8	1 5/8	1 11/16	1 3/4	1 7/8	1 15/16	2 1/16	2 1/8	2 1/4	2 5/16	2 7/16	2 1/2	2 5/8	2 3/4	2 7/8
1 1/2	1 3/4	1 13/16	1 7/8	1 15/16	2 1/16	2 1/8	2 3/16	2 5/16	2 3/8	2 1/2	2 5/8	2 11/16	2 13/16	2 15/16
1 5/8	1 7/8	1 15/16	2	2 1/16	2 1/8	2 3/16	2 5/16	2 3/8	2 1/2	2 9/16	2 11/16	2 3/4	2 7/8	3
1 3/4	1 15/16	2	2 1/16	2 1/8	2 1/4	2 5/16	2 3/8	2 1/2	2 9/16	2 11/16	2 3/4	2 7/8	2 15/16	3 1/16
1 7/8	2 1/16	2 1/8	2 3/16	2 1/4	2 5/16	2 3/8	2 1/2	2 9/16	2 5/8	2 3/4	2 13/16	2 15/16	3	3 1/8
2	2 3/16	2 1/4	2 5/16	2 3/8	2 7/16	2 1/2	2 9/16	2 11/16	2 3/4	2 13/16	2 15/16	3	3 1/8	3 3/16
2 1/8	2 5/16	2 3/8	2 3/8	2 7/16	2 1/2	2 5/8	2 11/16	2 3/4	2 13/16	2 15/16	3	3 1/8	3 3/16	3 1/4
2 1/4	2 7/16	2 7/16	2 1/2	2 9/16	2 5/8	2 11/16	2 3/4	2 7/8	2 15/16	3	3 1/8	3 3/16	3 1/4	3 3/8
2 3/8	2 1/2	2 9/16	2 5/8	2 11/16	2 3/4	2 13/16	2 7/8	2 15/16	3	3 1/8	3 3/16	3 1/4	3 3/8	3 7/16
2 1/2	2 5/8	2 11/16	2 3/4	2 13/16	2 7/8	2 15/16	3	3 1/16	3 1/8	3 3/16	3 1/4	3 3/8	3 7/16	3 9/16

NOTE: Values below or to right of upper zigzag lines are large enough for ¾″ rivets
" " " " " " lower " " " " " " ⅞″ "

Care should also be taken that the rivets are not so close that there will not be at least 1″ between the holes in the direction of the line of stress, see Fig. 43d.

In many cases a row of rivets must be driven below another row and in material which is perpendicular to the material in which the first row is driven. Such a case is in the cover plate of a plate girder, or for that matter in most cases of cover plates. In such cases it is desirable to know what spacing must be used in order that the dolly will not be interfered with by the rivet already driven in the other row. Table IX gives such information. It is to be noted that the value Y is the distance from the inner side of the leg of the angle, and is not the gauge. For example, let it be required to determine the minimum stagger for ¾-inch rivets in a 3½-inch leg of a $3\frac{1}{2}''\times3\frac{1}{2}''\times\frac{3}{8}''$ angle. The distance Y is then equal to the gauge of a 3-inch leg less the thickness of the angle, or

$$Y = 2'' - \tfrac{3}{8}''$$
$$= 1\tfrac{5}{8}''$$

TABLE IX

Minimum Staggers

$c = 1\frac{1}{8}''$ for $\frac{3}{4}''$ rivets $= 1\frac{1}{4}''$ " $\frac{7}{8}''$ "											

DIAMETER OF RIVETS	VALUES OF Y										
	$1\frac{1}{8}$	$1\frac{3}{16}$	$1\frac{1}{4}$	$1\frac{5}{16}$	$1\frac{3}{8}$	$1\frac{7}{16}$	$1\frac{1}{2}$	$1\frac{9}{16}$	$1\frac{5}{8}$	$1\frac{11}{16}$	$1\frac{3}{4}$
$\frac{3}{4}$	$1\frac{1}{4}$	$1\frac{3}{16}$	$1\frac{1}{8}$	$1\frac{1}{16}$	$\frac{15}{16}$	$\frac{7}{8}$	$\frac{3}{4}$	$\frac{5}{8}$	$\frac{3}{8}$	0	0
$\frac{7}{8}$	$1\frac{3}{8}$	$1\frac{5}{16}$	$1\frac{1}{4}$	$1\frac{3}{16}$	$1\frac{1}{8}$	1	$\frac{7}{8}$	$\frac{13}{16}$	$\frac{11}{16}$	$\frac{1}{2}$	$\frac{5}{16}$

All dimensions in inches

Looking along the top row the value $1\frac{5}{8}$ inches is found and going downward to the $\frac{3}{4}$-inch line of values, $\frac{3}{8}$ inch is found to be the least distance that the rivet under consideration may be driven from the one in the other leg of the angle.

In some cases it is possible to drive rivets opposite if the proper row is driven first. Thus, in the $5'' \times 3\frac{1}{2}'' \times \frac{3}{8}''$ angle of Fig. 44, if $\frac{3}{4}$-inch rivets in the 5-inch leg were driven first, those in the 3-inch leg must stagger by $\frac{3}{8}$ inch, as figured above, Fig. 44a, but if the rivets in the 3-inch leg were driven first, the distance $Y = 3'' - \frac{3}{8}''$

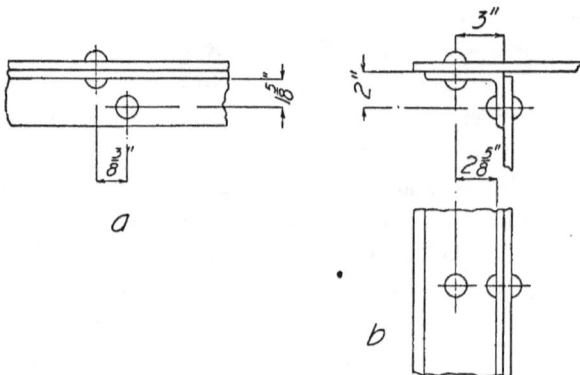

Fig. 44.   Rivet Stagger

$= 2\frac{5}{8}''$, which, being outside the values in Table IX, show that the rivets in the 5-inch leg may be driven with a zero stagger, or just opposite.

Certain clauses in most specifications call attention to the fact that rivets must not be used in tension. While it is desirable not to have rivets in tension, and their use to resist tensile stresses should not be encouraged, yet a rivet has a distinct value when used in tension. Also, tests of a confidential nature have come under the author's observation, and they tend to prove that rivets so used show as great an efficiency as a turned bolt of the same diameter.

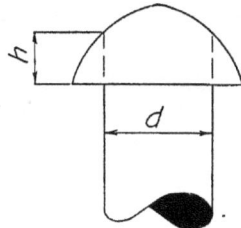

Fig. 45. Drawing of Standard Rivet

However, the strength of such rivets must not be assumed as being equal to a bolt of equal diameter, but must be computed. The head of the rivet must be drawn out to full size, and the distance "h," Fig. 45, determined. The value of the rivet in tension is then given by the formula

$$S_t = 3.14 \ S_s dh$$

where S_s = the unit shearing stress; d = the diameter of the rivet; and h = the value as determined above.

For a $\frac{7}{8}$-inch rivet $h = 0.45$ inch and, therefore, this value of the rivet in tension, S_s being taken at 10,000 pounds per square inch, is

$$S_t = 3.14 \times 10,000 \times \tfrac{7}{8} \times 0.45$$
$$= 12,360 \text{ pounds}$$

which is seen to be considerable, and which is equal to the body of the rivet being strained up to 20,050 pounds per square inch.

It is thus seen that the head more than develops the strength of the body of the rivet. Therefore, in figuring the amount a rivet should take in tension, one should multiply the area of the cross-section by the allowable unit stress decided upon. Since the specifications do not give this, it will be safe to use the ultimate strength for rivet steel with a factor of safety of 4. Since the ultimate strength of rivet steel should be about 50,000 pounds per square inch, this would make the allowable 12,500 pounds, and a $\frac{7}{8}$-inch rivet would have a value of

$$S_t = 12,500 \times 0.6013$$
$$= 7,510 \text{ pounds}$$

which is less than the amount required to strain its head up to the maximum allowable.

TABLE X

Rivet Spacing Multiplication Table

SPACES	PITCH IN INCHES															SPACES
	1⅛	1¼	1⅜	1½	1⅝	1¾	1⅞	2	2⅛	2¼	2⅜	2½	2⅝	2¾	2⅞	
1																1
2	- 2¼	- 2½	- 2¾	- 3	- 3¼	- 3½	- 3¾	- 4	- 4¼	- 4½	- 4¾	- 5	- 5¼	- 5½	- 5¾	2
3	- 3⅜	- 3¾	- 4⅛	- 4½	- 4⅞	- 5¼	- 5⅝	- 6	- 6⅜	- 6¾	- 7⅛	- 7½	- 7⅞	- 8¼	- 8⅝	3
4	- 4½	- 5	- 5½	- 6	- 6½	- 7	- 7½	- 8	- 8½	- 9	- 9½	-10	-10½	-11	-11½	4
5	- 5⅝	- 6¼	- 6⅞	- 7½	- 8⅛	- 8¾	- 9⅜	-10	-10⅝	-11¼	-11⅞	1- 0½	1- 1⅛	1- 1¾	1- 2⅜	5
6	- 6¾	- 7½	- 8¼	- 9	- 9¾	-10½	-11¼	1- 0	1- 0¾	1- 1½	1- 2¼	1- 3	1- 3¾	1- 4½	1- 5¼	6
7	- 7⅞	- 8¾	- 9⅝	-10½	-11⅜	1- 0¼	1- 1⅛	1- 2	1- 2⅞	1- 3¾	1- 4⅝	1- 5½	1- 6⅜	1- 7¼	1- 8⅛	7
8	- 9	-10	-11	1- 0	1- 1	1- 2	1- 3	1- 4	1- 5	1- 6	1- 7	1- 8	1- 9	1-10	1-11	8
9	-10⅛	-11¼	1- 0⅜	1- 1½	1- 2⅝	1- 3¾	1- 4⅞	1- 6	1- 7⅛	1- 8¼	1- 9⅜	1-10½	1-11⅝	2- 0¾	2- 1⅞	9
10	-11¼	1 0½	1- 1¾	1- 3	1- 4¼	1- 5½	1- 6¾	1- 8	1- 9¼	1-10½	1-11¾	2- 1	2- 2¼	2- 3½	2- 4¾	10
11	1- 0⅜	1- 1¾	1- 3⅛	1- 4½	1- 5⅞	1- 7¼	1- 8⅝	1-10	1-11⅜	2- 0¾	2- 2⅛	2- 3½	2- 4⅞	2- 6¼	2- 7⅝	11
12	1- 1½	1- 3	1- 4½	1- 6	1- 7½	1- 9	1-10½	2- 0	2- 1½	2- 3	2- 4½	2- 6	2- 7½	2- 9	2-10½	12
13	1- 2⅝	1- 4¼	1- 5⅞	1- 7½	1- 9⅛	1-10¾	2- 0⅜	2- 2	2- 3⅝	2- 5¼	2- 6⅞	2- 8½	2-10⅛	2-11¾	3- 1⅜	13
14	1- 3¾	1- 5½	1- 7¼	1- 9	1-10¾	2- 0½	2- 2¼	2- 4	2- 5¾	2- 7½	2- 9¼	2-11	3- 0¾	3- 2½	3- 4¼	14
15	1- 4⅞	1- 6¾	1- 8⅝	1-10½	2- 0⅜	2- 2¼	2- 4⅛	2- 6	2- 7⅞	2- 9¾	2-11⅝	3- 1½	3- 3⅜	3- 5¼	3- 7⅛	15
16	1- 6	1- 8	1-10	2- 0	2- 2	2- 4	2- 6	2- 8	2-10	3- 0	3- 2	3- 4	3- 6	3- 8	3-10	16
17	1- 7⅛	1- 9¼	1-11⅜	2- 1½	2- 3⅝	2- 5¾	2- 7⅞	2-10	3- 0⅛	3- 2¼	3- 4⅜	3- 6½	3- 8⅝	3-10¾	4- 0⅞	17
18	1- 8¼	1-10½	2- 0¾	2- 3	2- 5¼	2- 7½	2- 9¾	3- 0	3- 2¼	3- 4½	3- 6¾	3- 9	3-11¼	4- 1½	4- 3¾	18
19	1- 9⅜	1-11¾	2- 2⅛	2- 4½	2- 6⅞	2- 9¼	2-11⅝	3- 2	3- 4⅜	3- 6¾	3- 9⅛	3-11½	4- 1⅞	4- 4¼	4- 6⅝	19
20	1-10½	2- 1	2- 3½	2- 6	2- 8½	2-11	3- 1½	3- 4	3- 6½	3- 9	3-11½	4- 2	4- 4½	4- 7	4- 9½	20
21	1-11⅝	2- 2¼	2- 4⅞	2- 7½	2-10⅛	3- 0¾	3- 3⅜	3- 6	3- 8⅝	3-11¼	4- 1⅞	4- 4½	4- 7⅛	4- 9¾	5- 0⅜	21
22	2- 0¾	2- 3½	2- 6¼	2- 9	2-11¾	3- 2½	3- 5¼	3- 8	3-10¾	4- 1½	4- 4¼	4- 7	4- 9¾	5- 0½	5- 3¼	22
23	2- 1⅞	2- 4¾	2- 7⅝	2-10½	3- 1⅜	3- 4¼	3- 7⅛	3-10	4- 0⅞	4- 3¾	4- 6⅝	4- 9½	5- 0⅜	5- 3¼	5- 6⅛	23
24	2- 3	2- 6	2- 9	3- 0	3- 3	3- 6	3- 9	4- 0	4- 3	4- 6	4- 9	5- 0	5- 3	5- 6	5- 9	24
25	2- 4⅛	2- 7¼	2-10⅜	3- 1½	3- 4⅝	3- 7¾	3-10⅞	4- 2	4- 5⅛	4- 8¼	4-11⅜	5- 2½	5- 5⅝	5- 8¾	5-11⅞	25
26	2- 5¼	2- 8½	2-11¾	3- 3	3- 6¼	3- 9½	4- 0¾	4- 4	4- 7¼	4-10½	5- 1¾	5- 5	5- 8¼	5-11½	6- 2¾	26
27	2- 6⅜	2- 9¾	3- 1⅛	3- 4½	3- 7⅞	3-11¼	4- 2⅝	4- 6	4- 9⅜	5- 0¾	5- 4⅛	5- 7½	5-10⅞	6- 2¼	6- 5⅝	27
28	2- 7½	2-11	3- 2½	3- 6	3- 9½	4- 1	4- 4½	4- 8	4-11½	5- 3	5- 6½	5-10	6- 1½	6- 5	6- 8½	28
29	2- 8⅝	3- 0¼	3- 3⅞	3- 7½	3-11⅛	4- 2¾	4- 6⅜	4-10	5- 1⅝	5- 5¼	5- 8⅞	6- 0½	6- 4⅛	6- 7¾	6-11⅜	29
30	2- 9¾	3- 1½	3- 5¼	3- 9	4- 0¾	4- 4½	4- 8¼	5- 0	5- 3¾	5- 7½	5-11¼	6- 3	6- 6¾	6-10½	7- 2¼	30
SPACES	1⅛	1¼	1⅜	1½	1⅝	1¾	1⅞	2	2⅛	2¼	2⅜	2½	2⅝	2¾	2⅞	SPACES
	PITCH IN INCHES															

TABLE X (Continued)
Rivet Spacing Multiplication Table

SPACES	PITCH IN INCHES 3	3⅛	3¼	3⅜	3½	3¾	4	4¼	4½	4¾	5	5¼	5½	5¾	6	SPACES
1																1
2	·6	·6¼	·6½	·6¾	·7	·7½	·8	·8½	·9	·9½	·10	·10½	·11	·11½	1-0	2
3	·9	·9⅜	·9¾	·10⅛	·10½	·11¼	1-0	1-0¾	1-1½	1-2¼	1-3	1-3¾	1-4½	1-5¼	1-6	3
4	1-0	1-0½	1-1	1-1½	1-2	1-3	1-4	1-5	1-6	1-7	1-8	1-9	1-10	1-11	2-0	4
5	1-3	1-3⅝	1-4¼	1-4⅞	1-5½	1-6¾	1-8	1-9¼	1-10½	1-11¾	2-1	2-2¼	2-3½	2-4¾	2-6	5
6	1-6	1-6¾	1-7½	1-8¼	1-9	1-10½	2-0	2-1½	2-3	2-4½	2-6	2-7½	2-9	2-10½	3-0	6
7	1-9	1-9⅞	1-10¾	1-11⅝	2-0½	2-2¼	2-4	2-5¾	2-7½	2-9¼	2-11	3-0¾	3-2½	3-4¼	3-6	7
8	2-0	2-1	2-2	2-3	2-4	2-6	2-8	2-10	3-0	3-2	3-4	3-6	3-8	3-10	4-0	8
9	2-3	2-4⅛	2-5¼	2-6⅜	2-7½	2-9¾	3-0	3-2¼	3-4½	3-6¾	3-9	3-11¼	4-1½	4-3¾	4-6	9
10	2-6	2-7¼	2-8½	2-9¾	2-11	3-1½	3-4	3-6½	3-9	3-11½	4-2	4-4½	4-7	4-9½	5-0	10
11	2-9	2-10⅜	2-11¾	3-1⅛	3-2½	3-5¼	3-8	3-10¾	4-1½	4-4¼	4-7	4-9¾	5-0½	5-3¼	5-6	11
12	3-0	3-1½	3-3	3-4½	3-6	3-9	4-0	4-3	4-6	4-9	5-0	5-3	5-6	5-9	6-0	12
13	3-3	3-4⅝	3-6¼	3-7⅞	3-9½	4-0¾	4-4	4-7¼	4-10½	5-1¾	5-5	5-8¼	5-11½	6-2¾	6-6	13
14	3-6	3-7¾	3-9½	3-11¼	4-1	4-4½	4-8	4-11½	5-3	5-6½	5-10	6-1½	6-5	6-8½	7-0	14
15	3-9	3-10⅞	4-0¾	4-2⅝	4-4½	4-8¼	5-0	5-3¾	5-7½	5-11¼	6-3	6-6¾	6-10½	7-2¼	7-6	15
16	4-0	4-2	4-4	4-6	4-8	5-0	5-4	5-8	6-0	6-4	6-8	7-0	7-4	7-8	8-0	16
17	4-3	4-5⅛	4-7¼	4-9⅜	4-11½	5-3¾	5-8	6-0¼	6-4½	6-8¾	7-1	7-5¼	7-9½	8-1¾	8-6	17
18	4-6	4-8¼	4-10½	5-0¾	5-3	5-7½	6-0	6-4½	6-9	7-1½	7-6	7-10½	8-3	8-7½	9-0	18
19	4-9	4-11⅜	5-1¾	5-4⅛	5-6½	5-11¼	6-4	6-8¾	7-1½	7-6¼	7-11	8-3¾	8-8½	9-1¼	9-6	19
20	5-0	5-2½	5-5	5-7½	5-10	6-3	6-8	7-1	7-6	7-11	8-4	8-9	9-2	9-7	10-0	20
21	5-3	5-5⅝	5-8¼	5-10⅞	6-1½	6-6¾	7-0	7-5¼	7-10½	8-3¾	8-9	9-2¼	9-7½	10-0¾	10-6	21
22	5-6	5-8¾	5-11½	6-2¼	6-5	6-10½	7-4	7-9½	8-3	8-8½	9-2	9-7½	10-1	10-6½	11-0	22
23	5-9	5-11⅞	6-2¾	6-5⅝	6-8½	7-2¼	7-8	8-1¾	8-7½	9-1¼	9-7	10-0¾	10-6½	11-0¼	11-6	23
24	6-0	6-3	6-6	6-9	7-0	7-6	8-0	8-6	9-0	9-6	10-0	10-6	11-0	11-6	12-0	24
25	6-3	6-6⅛	6-9¼	7-0⅜	7-3½	7-9¾	8-4	8-10¼	9-4½	9-10¾	10-5	10-11¼	11-5½	11-11¾	12-6	25
26	6-6	6-9¼	7-0½	7-3¾	7-7	8-1½	8-8	9-2½	9-9	10-3½	10-10	11-4½	11-11	12-5½	13-0	26
27	6-9	7-0⅜	7-3¾	7-7⅛	7-10½	8-5¼	9-0	9-6¾	10-1½	10-8¼	11-3	11-9¾	12-4½	12-11¼	13-6	27
28	7-0	7-3½	7-7	7-10½	8-2	8-9	9-4	9-11	10-6	11-1	11-8	12-3	12-10	13-5	14-0	28
29	7-3	7-6⅝	7-10¼	8-1⅞	8-5½	9-0¾	9-8	10-3¼	10-10½	11-5¾	12-1	12-8¼	13-3½	13-10¾	14-6	29
30	7-6	7-9¾	8-1½	8-5¼	8-9	9-4½	10-0	10-7½	11-3	11-10½	12-6	13-1½	13-9	14-4½	15-0	30
SPACES	3	3⅛	3¼	3⅜	3½	3¾	4	4¼	4½	4¾	5	5¼	5½	5¾	6	SPACES
	PITCH IN INCHES															

On account of the fact that riveted heads are not driven symmetrically the value of the rivet in tension is not certain, and their use in tension is not to be advised. Use turned bolts.

The bearing and shearing values of rivets may be found in the handbooks of the various manufacturers. The values for all values of allowable stresses are not usually given, but by a little trouble almost any values may be obtained by dividing those values there given, or by taking a multiple of them. For example, the bearing value of a $\frac{7}{8}$-inch rivet in a $\frac{7}{16}$-inch plate, unit allowable bearing stress 18,000 pounds, may be obtained by taking $1\frac{1}{2}$ times the value given in the 12,000-pound table, giving 6,885 pounds.

In cases of the webs of channels or I-beams, or other thicknesses of metal which are not in even sixteenths of an inch, but are given in decimal fractions, the values may, with the help of the slide rule, be obtained from the tables. For example, let it be required to find the value of a $\frac{7}{8}$-inch rivet in bearing in the web of a $15''\times33\#$ channel, the unit-bearing stress allowed being 15,000 pounds. From Cambria, the thickness is seen to be 0.4, and the bearing of a $\frac{7}{8}$-inch rivet in a $\frac{1}{2}$-inch plate is found to be 6,563 pounds. Therefore, the value sought will be

$$V = \frac{6,563}{0.5} \times 0.4 = 5,250 \text{ pounds}$$

For convenience in rivet spacing, Table X will be found convenient, the value of any number of spaces of a given length being determined at a glance.

Bolts, Nuts, and Washers. Bolts are made by forming a head on one end and cutting a thread on the other end of an iron rod. In such cases the body of the bolt does not represent the strength, but the area at the root of the threads. In the handbooks is given the diameter of the screw thread for any bar or bolt of given diameter, and from this the strength of a bolt may be calculated, once the allowable unit tensile stress is determined. The diameter given for the rod or bolt is the diameter of the upset screw end. The strength of the bolt is then obtained by multiplying the diameter of the screw at root of thread by itself, by 0.7854,* and by the allowable unit stress, thus

$$\text{Strength of Bolt} = 0.7854 \ d_1{}^2 \times S$$

*In some books the area at root of thread is given direct.

TABLE XI
Standard Cast O. G. Washers

DIMENSIONS

$a = 4d + \frac{1}{4}''$

$b = 2d + \frac{1}{4}''$

$t = d$

$c = d + \frac{1}{8}''$

DIAMETER OF BOLT d	$\frac{1}{2}$	$\frac{5}{8}$	$\frac{3}{4}$	$\frac{7}{8}$	1	$1\frac{1}{8}$	$1\frac{1}{4}$	$1\frac{3}{8}$	$1\frac{1}{2}$	$1\frac{3}{4}$	2
DIAMETER a	$2\frac{1}{4}$	$2\frac{3}{4}$	$3\frac{1}{4}$	$3\frac{3}{4}$	$4\frac{1}{4}$	$4\frac{3}{4}$	$5\frac{1}{4}$	$5\frac{3}{4}$	$6\frac{1}{4}$	7	$8\frac{1}{4}$
Wt. OF 100 WASHERS IN Lbs.	21	43	70	113	175	256	332	455	610	865	1115
All dimensions in inches											

For example, let it be required to determine the strength of a 1½-inch bolt, the unit allowable stress in tension being 18,000 pounds per square inch. It is

$$\text{Strength of Bolt} = 0.7854 \times 1.284^2 \times 18,000$$
$$= 23,350 \text{ pounds}$$

while if the area of the body of the bolt was used the strength would be 31,800, from which it is seen that in determining the strength of bolts care must be taken to use the diameter at the root of the thread.

Information regarding bolts and nuts in general is given in the handbooks. Here the exact dimensions of the heads and nuts are given. In detailing it will be sufficiently accurate to assume the side of a square head or nut or the short diameter of a hexagonal head or nut as twice the diameter of the bolt, the thickness of each being equal to the diameter of the bolt.

When the nut is screwed up, the bolt should extend from ⅛ inch to ¼ inch above the nut.

Washers are of two kinds, cast and cut. The former are designated as O. G. (pronounced Oh Gee) washers on account of the curve given to their side. The sizes and weights of O. G. washers are given in Table XI.

Cut washers are made by stamping them out of sheet metal, and are principally used as separators where two angles are bolted together, or under the heads and nuts of small bolts which bolt timber in place. General information regarding them is given in Table XII.

TABLE XII
Standard Cut Washers
(In 200-pound kegs)

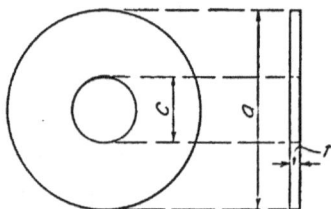

$$a = 2d + \tfrac{1}{2}''$$
$$c = d + \tfrac{1}{16}'' \text{ up to } d = 1''$$
$$= d + \tfrac{1}{8}'' \text{ when } d \text{ is greater than } 1''$$

SIZE OF BOLT OR UPSET = d	DIAMETER a	DIAMETER OF HOLE c	THICKNESS t	NO IN 100 POUNDS
3/16	5/8	1/4	3/64	36 200
1/4	3/4	5/16	1/16	4 900
5/16	7/8	3/8	•	11 100
3/8	1	7/16	5/64	6 700
7/16	1 1/4	1/2	"	4 100
1/2	1 3/8	9/16	7/64	2 600
5/8	1 1/2	5/8	"	2 700
5/8	1 3/4	11/16	9/64	1 300
3/4	2	13/16	"	1 000
7/8	2 1/4	15/16	5/32	728
1	2 1/2	1 1/16	3/32	595
1 1/8	2 3/4	1 1/4	"	507
1 1/4	3	1 3/8	"	428
1 3/8	3 1/4	1 1/2	11/64	328
1 1/2	3 1/2	1 5/8	"	284
1 5/8	3 3/4	1 3/4	11/64	248
1 3/4	4	1 1/4	"	218
1 7/8	4 1/4	2	"	194
2	4 1/2	2 1/8	"	173
All dimensions in inches				

Tension Members. These may consist of square, round, or rectangular bars, or they may be of shapes riveted together. The latter class will be considered under the detailing of tension members.

When the bar is square or circular in section, it may be formed into loops at its ends, or upset and nuts put on, in order to attach it to other parts of the structure in which it is used. In the former case it is called a loop bar. In case it is rectangular in section it may be formed into a loop bar, or may have its ends forged out into a somewhat circular shape, see Fig. 46, and a hole bored in them in order to connect them to the rest of the structure. In this case it is called an eye bar.

In order to be assured that the eye bar will not break in the head, the distances a are made such that $2a$ is greater than w, usually between $1.3w$ and $1.4w$. If not required by the specifications, it is usually left to the manufacturers with the stipulation that the eye bars must break in the body of the bar, not in the head.

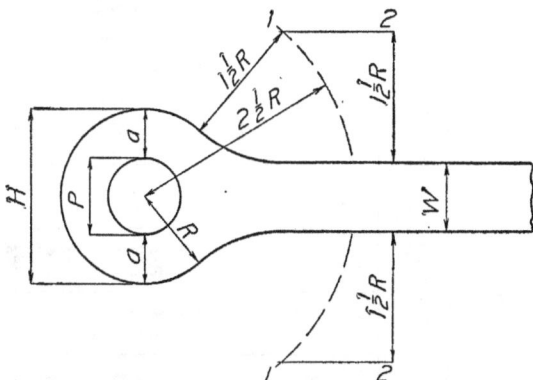

Fig. 46. Dimensions Required in Eye-Bar Design

The dimensions of eye bars are given in the handbooks. In Cambria the excess through the pin hole for the 2-inch bar is $(4\frac{1}{2} - 1\frac{7}{8})$ $\div 2 = 1\frac{5}{16}$, an excess of 33 per cent.

Care should be taken to note that the values here given are the minimum thicknesses. Bars thinner than these are liable to upset so imperfectly as to be unsafe in the heads. An eye bar should not, as a rule, be less in thickness than one-sixth of the depth. The pins given in the tables in the handbooks are maximum pins. The

TABLE XIII

Loop Bars. Allowance for One Loop. American Bridge Company Standards

(All dimensions in inches)

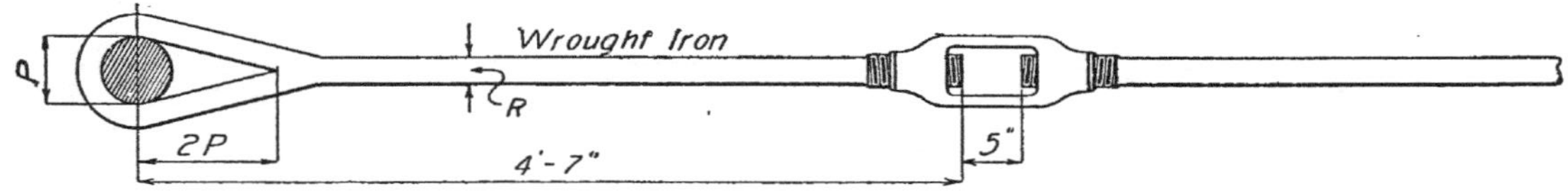

Length in inches beyond pin center to form one eye equals 3.7 (P + R) + 5"

DIAM. OF PINS	DIAMETER OR SIDE OF BARS ¾	⅞	1	1⅛	1¼	1⅜	1½	1⅝	1¾	1⅞	2	2⅛	2¼	2⅜	2½	2⅝	2¾	2⅞	3	DIAM. OF PINS
1	11½	12	12½																	1
1¼	12½	12⅞	13⅜	13⅞	14¼															1¼
1½	13⅜	13⅞	14¼	14¾	15¼	15⅝	16⅛													1½
1¾	14¼	14¾	15¼	15⅝	16⅛	16⅝	17	17½	18											1¾
2	15¼	15⅝	16⅛	16⅝	17	17½	18	18⅜	18⅞	19⅜	19¾									1¾
2¼	16⅛	16⅝	17	17½	18	18⅜	18⅞	19⅜	19¾	20¼	20¾	21⅛	21⅜							2
2½	17	17½	18	18⅜	18⅞	19⅜	19¾	20¼	20¾	21⅛	21⅝	22⅛	22⅝	23	23½					2¼
2¾	18	18⅜	18⅞	19⅜	19¾	20¼	20¾	21⅛	21⅝	22⅛	22⅝	23	23½	24	24⅜	24⅞	25⅜			2½
3	18⅞	19⅜	19¾	20¼	20¾	21⅛	21⅝	22⅛	22⅝	23	23½	24	24⅜	24⅞	25⅜	25⅞	26¼	26¾	27¼	2¾
3¼	19¾	20¼	20¾	21⅛	21⅝	22⅛	22⅝	23	23½	24	24⅜	24⅞	25⅜	25⅞	26¼	26¾	27¼	27⅝	28⅛	3
3½	20¾	21⅛	21⅝	22⅛	22⅝	23	23½	24	24⅜	24⅞	25⅜	25⅞	26¼	26¾	27¼	27⅝	28⅛	28⅜	29	3¼
3¾	21⅝	22⅛	22⅝	23	23½	24	24⅜	24⅞	25⅜	25⅞	26¼	26¾	27¼	27⅝	28⅛	28⅝	29	29½	30	3½
4	22⅝	23	23½	24	24⅜	24⅞	25⅜	25⅞	26¼	26¾	27¼	27⅝	28⅛	28⅝	29	29½	30	30½	31	3¾
4¼	23½	24	24⅜	24⅞	25⅜	25⅞	26¼	26¾	27¼	27⅝	28⅛	28⅝	29	29½	30	30½	31	31⅜	31⅞	4
4½	24⅜	24⅞	25⅜	25⅞	26¼	26¾	27¼	27⅝	28⅛	28⅜	29	29½	30	30½	31	31⅜	31⅞	32¼	32¾	4¼
4¾	25⅜	25⅞	26¼	26¾	27¼	27⅝	28⅛	28⅝	29	29½	30	30½	31	31⅜	31⅞	32¼	32¾	33¼	33⅝	4½
5	26¼	26¾	27¼	27⅝	28⅛	28⅝	29	29½	30	30½	31	31⅜	31⅞	32¼	32¾	33¼	33⅝	34⅛	34⅝	4¾
5¼	27¼	27⅝	28⅛	28⅝	29	29½	30	30½	31	31⅜	31⅞	32¼	32¾	33¼	33⅝	34⅛	34⅝	35	35½	5
5½	28⅛	28⅝	29	29½	30	30½	31	31⅜	31⅞	32¼	32¾	33¼	33⅝	34⅛	34⅝	35	35½	36	36½	5¼
5¾	29	29½	30	30½	31	31⅜	31⅞	32¼	32¾	33¼	33⅝	34⅛	34⅝	35	35	36	36½	37	37⅜	5½
6	30	30½	31	31⅜	31⅞	32¼	32¾	33¼	33⅝	34⅛	34⅝	35	35½	36	36½	37	37⅜	37⅞	38⅜	6

NOTE: Maximum shipping length should not exceed 35 feet   All dimensions in inches

American Bridge Company practice requires the smallest pin to be not less than three-fourths the width of the eye bar.

Bars of a square or circular section could, as in the case of bolts, have a screw thread cut on their ends and by means of nuts be connected to the other part of the structure, but such an operation would be costly since the bars are long and much of the section would be wasted for a great length. In such cases the bars are ordered 6 inches longer than required and this 6 inches is, after heating to a welding heat, upset or pushed in 6 inches, thus increasing the diameter of the bar at the end so that the diameter at the bottom of the screw threads will be greater than the diameter of the original bar. This is done so that the bar will break in the body, and not at the joint.

The sizes of upsets for bars of various sizes are given in the handbooks. Let it be required to determine the size hole through which a $1\frac{1}{8}$-inch bar with upset end would pass and the nut required. We find opposite the $1\frac{1}{8}$ the value $1\frac{1}{2}$, showing that the upset will be $1\frac{1}{2}$ inches. In another table opposite $1\frac{1}{2}$ is given the size and weight of a square nut, viz, $1\frac{1}{2}$ inches thick, 3 inches on the side, and weight 3.175 pounds. The use of square nuts is not to be encouraged, the hexagonal form being the better, on account of their lighter weight.

Instead of the rods being fitted with nuts and threads at their ends, they may, as mentioned above, be made into loop bars. Loop bars are welded, and for this reason are not to be desired since welds are never as strong as the original. However, the loop bar has 100 per cent excess through the pin, and in order to have an efficiency of 100 per cent it must have a weld with an efficiency of 50 per cent. Since such a weld is well within the limits of possibility, it is permissible to use loop bars in highway bridges or other structures where the impact is not great, and in counters, since here the pins are usually of such a diameter that they would be too great for an eye bar of the section of the counter. Table XIII gives information regarding loop bars. They must be made of wrought iron since steel does not weld well.

Clearances. It is very important that each member of a structure fit together well in the field; and it is equally important that the draftsman should so detail his work that the various parts of any particular member should, without further cutting than the first, fit together. Also the rivets should be so spaced and placed that they can be driven.

The rivet clearances have been mentioned under "Rivets and Rivet Spacing" and will not be taken up here. It is sufficient to say that on the rivet clearances is where the novice makes the most of his mistakes.

Where the distance between the outer faces of several members placed together is to be computed, it is necessary, on account of the liability of plates to exceed their nominal thicknesses, and rivet heads their nominal height, to make certain allowances. The usual practice is:

(1) Between eye (or loop) bars allow $\frac{1}{16}$ inch.
(2) Between an eye(or loop) bar and a built-up member $\frac{1}{8}$ inch.
(3) Between two built-up members $\frac{1}{4}$ inch.

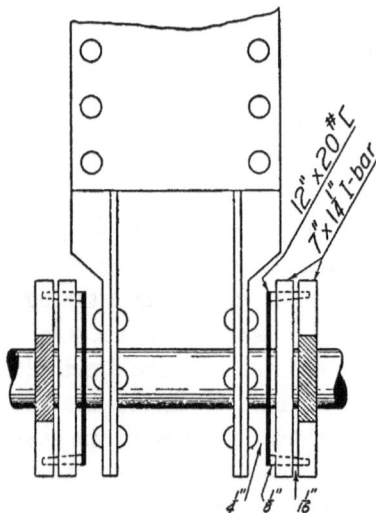

Fig. 47. Joint Showing Clearance between Members

For example, suppose it was required to compute the distance out to out of the members shown in Fig. 47. The clearance would be as indicated, and the distance D would be:

$$D = 2\left(\frac{10}{2} + 0.4 + \tfrac{5}{8} + \tfrac{1}{4} + 0.28 + \tfrac{1}{8} + 1\tfrac{1}{4} + \frac{1}{16} + 1\tfrac{1}{4}\right)$$
$$= 18.485 = 18\tfrac{1}{2} \text{ inches}$$

This value would be the grip of the pin which was used at this joint. The 0.4 inch and 0.28 inch in the above are the thicknesses of the channel webs, and the $\frac{5}{8}$ inch is the height of a $\frac{7}{8}$ inch rivet head.

In the use of eye bars, it is essential to see that their heads as well as their bodies clear. In order to determine the dimension of a section for the necessary clearance, the size of the head must be ascertained. This is best done by drawing up the head to a large scale. The method of procedure is as follows: (1) Draw the circle representing the pinhole; (2) for the width of eye bar under consideration, subtract the radius of the largest pinhole in Cambria for that bar from the radius of the given head and add the result to one-half the pinhole diameter in your particular case, thus giving you R, Fig. 46; (3) with the radius R describe a full circle; (4) with

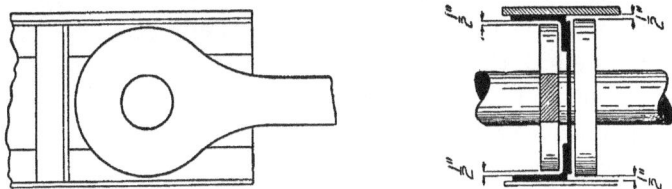

Fig. 48.   Eye-Bar and Built-Up Member Showing Clearance Allowed

the center of the pin as a center and a radius equal to $2\frac{1}{2}$ R describe a couple of arcs 1, 1; (5) parallel to the bar and at a distance $1\frac{1}{2}$ R from it, draw two lines, 2, 2, intersecting the arcs 1, 1; and (6) with these intersections as centers and a radius equal to $1\frac{1}{2}$ R describe the small arcs completing the head, see Fig. 46.

No material should be closer to the edge of the eye-bar head than $\frac{1}{2}$ inch.   This clearance should always be given, see Fig. 48,

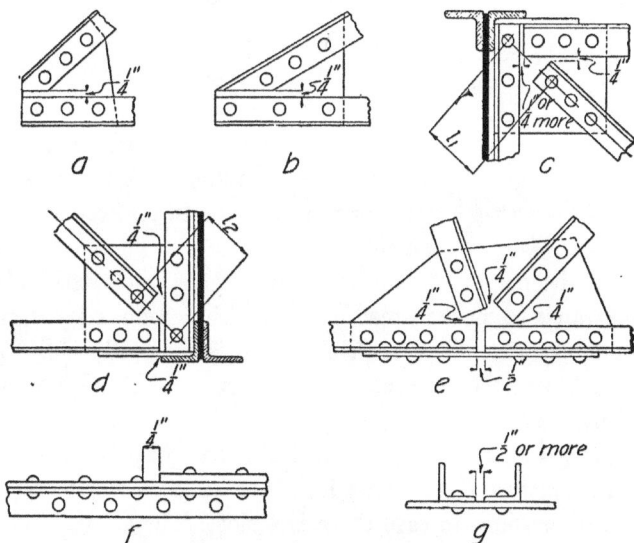

Fig. 49.   Riveted Joints Showing Clearance Allowed

although the clearance of $\frac{1}{8}$ or $\frac{1}{16}$ on the side should be allowed as usual in case it was against a built-up member or another eye bar.

In case the head is on the interior of a channel or so as to come near the fillet of an angle, the $\frac{1}{2}$ inch must be measured from the curve of the fillet. This $\frac{1}{2}$ inch does not apply to the body of the bar, the clearance there being $\frac{1}{4}$ inch in accordance with what follows.

Wherever several pieces of metal are riveted to the same side of a plate or other member and could, theoretically, come close against each other, $\frac{1}{4}$-inch clearance is allowed for each case where the ends are not planed. This allows for the slight variations in length liable to occur when the surfaces are sheared. The members will then be sufficiently close together for all practical purposes. In order that no errors occur, the joint should be drawn up on a separate sheet to a scale of at least $1\frac{1}{2}$ or 2 inches to the foot in case the pieces meet at an angle. In case the pieces meet at right angles, the distances may be computed. Fig. 49 gives a few of the most common cases.

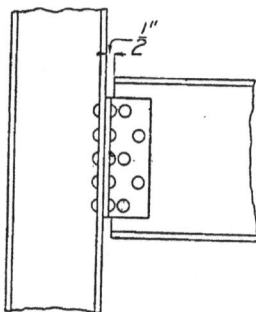

Fig. 50. Column and Beam Connection Showing Clearance

As in the case of Fig. 49c and 49d the clearances at one end will be $\frac{1}{4}$ inch and at the other end may be more, and should be, in order that the distances l_1 and l_2 shall be the same. (The distance from the first rivet to the end of the angle is usually $1\frac{1}{4}$ or $1\frac{1}{2}$, generally the latter.) It must not be understood that the clearance is exactly $\frac{1}{4}$ inch; it must be at least $\frac{1}{4}$ inch, and may be more, up to $\frac{3}{8}$ inch or $\frac{1}{2}$ inch in order that the distance from the rivet to some other point or rivet may be in an even $\frac{1}{16}$ inch or $\frac{1}{8}$ inch.

When I-beams or channels are placed as mentioned above, $\frac{1}{2}$-inch clearance or more instead of the $\frac{1}{4}$-inch is required, one of the most common cases where such clearance is required being shown in Fig. 50. For other clearances in beams see "The Detailing of Beams," page 72.

Wherever bolts, rods, upsets, or rolled bars pass through a hole or slot, the aperture should be $\frac{1}{8}$ inch greater in diameter or $\frac{1}{4}$ inch greater in dimensions in case there is a slot. The above is in case the material is rolled steel or iron. In case of a casting, $\frac{1}{4}$ inch should be added to the dimensions of the member which is to pass through the opening.

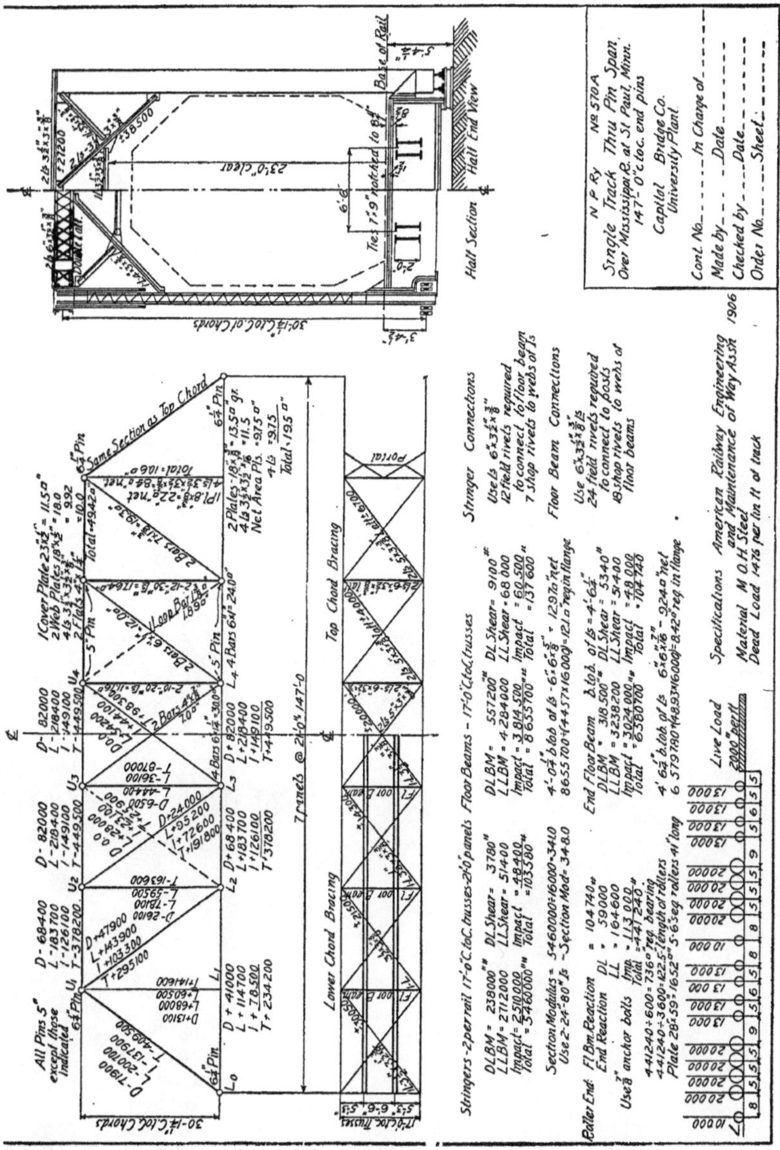

Plate I. Stress Sheet of Truss Bridge

LA SALLE STATION, L. S. & M. S. AND C., R. I. & P. RAILROADS, CHICAGO

STRUCTURAL DRAFTING

PART II

DETAILING METHODS

Detailing of Angles. The line upon which the rivets are spaced is the gauge line. The standard gauges given in Table VII should not be departed from unless instructions are given otherwise or unless it is impossible to make the detail without doing so, in which cases a "special" gauge is used. In deciding upon a special gauge care must be taken to see that the gauge line is not less than the standard edge and clearance distances for the rivet used. In Fig. 51 is shown the minimum values of these distances.

In some cases, as in the flange of plate girders, the rivet spacing is determinate, but in the majority of cases this is not so. In the former case the spacing between certain limits should be made equal to that at the lower limit. It is unwise and not economical to change the spacing every few feet. When the spacing is not determined, the rivets may be placed as desired, the only limitations being (1) that they can be driven, and (2) that they do not take out too much section, providing that the angle is in tension. Of course the limitations as to maximum and minimum spacings apply here, the spacing, being used from 3 inches to 4 or $4\frac{1}{2}$ inches for $\frac{3}{4}$-inch or $\frac{7}{8}$-inch rivets, the lower limit being used if possible in order to keep down the size of the connection plates. It is economical to make all the spaces equal. The spacing may be governed by the desire to have the connection plate symmetrical. The distance from the end of the angle to the first rivet is usually $1\frac{1}{2}$ inches for $\frac{7}{8}$-inch rivets, but 2 inches is sometimes, though seldom, used. In case two angles are used as tension

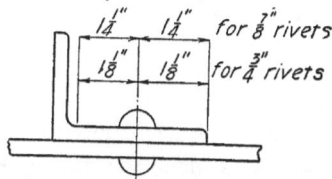

Fig. 51.   Minimum Clearance Distances in Angle Detailing

members, they may be riveted together at distances not greater than 12 to 18 inches.

The gauge line in single gauge angles is used as the working line, and passes through the working points. In the case of a double-gauge line, the inner gauge line should be used as the line of reference since by so doing the stresses in the angle will be less than if the reference line was taken midway between the two lines.

Angles in either tension or compression should be connected by both legs, otherwise the stresses due to eccentricity will cause the total stresses to be far above the average stress, as a usual case 100 per cent. This should be done by a "clip" angle, and as many rivets should go from the angle into the clip as go from the clip into the connection plate. As a usual thing it is not necessary to detail

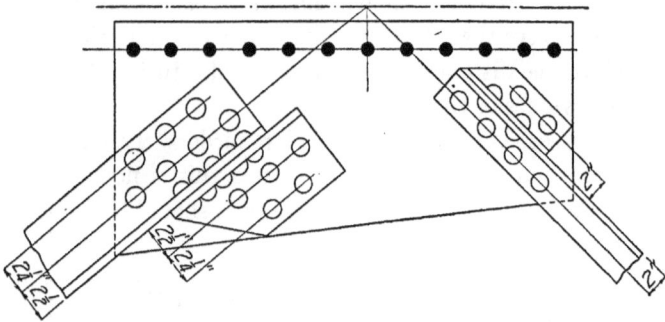

Fig. 52. Angle Detailing—Vertical Leg Not Shown

the vertical leg of the clip, as the shopmen will attend to it; only show the heads of the rivets. Figs. 50 and 52 illustrate the principles mentioned above.

The gauge should always be placed on the angle whether it is standard or special, but do not give the distances from the gauge to the edge of the leg.

In detailing diagonals, the end distances to the working points should be so chosen that the length of the angle will end in an eighth of an inch. If anything, make the angle a little short center to center of end holes to accomplish this. The member can easily be drawn up in place by the moderate use of drift pins, this making it taut when riveted up in place.

Detailing of Plates. The governing features of the detailing of plates are: (1) To keep the plate as small as can consistently be done; (2) to cut it as few times as possible in getting it into the final shape; (3) to have it symmetrical if possible; (4) to keep it an even number of inches in width; (5) to make it thick enough so that the number of rivets will be small and, therefore, the plate also; and (6) to detail it so that the rivet centers may be determined quickly and with certainty.

The plate is kept as small as possible for economical reasons, and for the same reason it should be cut as few times as possible, two cuts being the maximum and the desired number. It is usually more economical to leave the material on the plate than to trim it up. Therefore, it is important that the rivet spacing be so arranged that this can be done. In such cases the company is not only saved the labor of trimming the plate into some irregular shape, but it gets paid for the extra weight left on. Fig. 53a illustrates a plate poorly

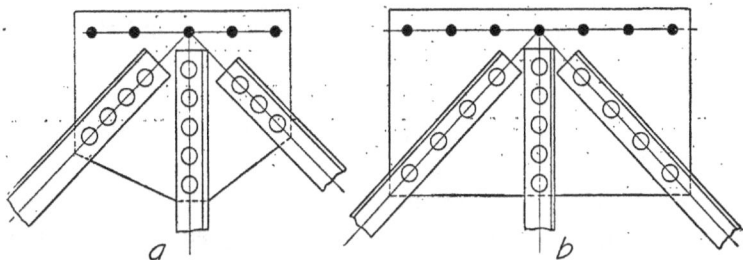

Fig. 53. Methods of Detailing Angles. (a) Poor Detailing. (b) Good Detailing

detailed and Fig. 53b, one well detailed. In the latter the plate, although somewhat larger, is of rectangular shape, and most engineers would prefer it to the other.

It is not always possible to have a plate symmetrical. When a plate is symmetrical, the templet work, and, therefore, the cost, is much reduced. If they cannot be made symmetrical, the next best thing is to have as many as possible alike or of the same width. With a little thought along these lines a draftsman can save his week's wages for his company many a day.

For economy's sake the plate should be in an even number of inches in width. Plates are not rolled in fractions of an inch except below 6 inches. If a plate is so detailed as to require a fraction of an inch in width, the next wider plate must be ordered and cut down to the required size, thus causing expense due to metal wasted and to labor required to get it cut to size. The length may be any dimension, but it is best to have it in even eighths of an inch, thus $2'-8\frac{1}{2}''$ or $3'-5\frac{7}{8}''$, etc. As mentioned before, the width of the plate must be stated in inches, not feet and inches. A plate is noted thus 1-Pl. $18''\times\frac{3}{8}''\times2'-5\frac{1}{8}''$ or 1-Pl. $25''\times\frac{3}{8}''\times3'-8\frac{1}{4}''$.

If the stress in a member is great, the number of rivets will necessarily be large. In such cases the thickness of the plate may be made thicker than $\frac{3}{8}$ inch in railroad, or $\frac{5}{16}$ inch in highway or building work, and the member so arranged as to bring the rivets in double shear. Unless the rivets can be brought in double shear it is unnecessary to increase the thickness of the plate, for shear governs in case of a $\frac{3}{8}$-inch plate, and in case of a $\frac{5}{16}$-inch plate the change to a $\frac{3}{8}$-inch plate does not reduce the number of rivets sufficiently to warrant it.

A plate must also be of sufficient section to transmit the stress from one member to the other. Since the area between the rivets is greater than that bearing upon the rivet, this is automatically attended to. In the case of shear along planes between members a different condition obtains. Here the computations are more or less involved, but the draftsman need not consider this phase of the design since in such cases the experienced designer will design the plate and it will come to him with the correct thickness marked upon it.

The rectangular detailing of plates is, of course, simple. The rivets are, except in such cases as in the webs of plate girders, spaced after the manner of the spacing in angles, the same conditions governing. The lines on which the rivets are placed are in such cases parallel to one of the sides of the plate and the spacing is readily laid off.

When a diagonal row of rivets is on a plate, it may be detailed in two ways: (1) By rectangular co-ordinates; and (2) by spacing along a line located by a bevel. Only in exceptional cases is the first method, shown in Fig. 54a, to be used.

The better way, and indeed it might be said the standard way, is shown in Fig. 54b. This way is easier for all concerned, and, in case of diagonals, lends itself especially well, since the distance between working points of the plates at the ends can be computed, and the rivet spacing being measured along this line gives the distance between the last holes on the plates by simple subtraction.

A plate should be detailed from one edge (the working edge) and the working point. The various distances to the rivet holes are measured from these places of reference. The distance from the last rivet hole to the far or side edges is not given. The plate of course being originally laid out to a large scale—a layout—care is taken that if the distances measured out from the working edge or point are used, the last rivet will not come closer to or farther from the far side of the plate than is allowed by specifications. In the

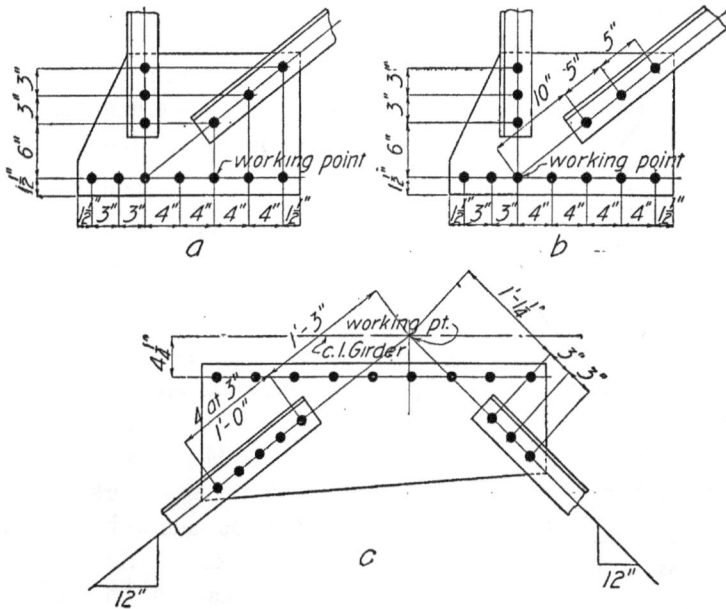

Fig. 54. Methods of Detailing Diagonal Row of Rivets. (a) Poor. (b) Good. (c) Example Where Working Point is Not on the Plate

case of the $\frac{7}{8}$-inch rivet, these limits are $1\frac{1}{2}$ inches for the smaller and 2 inches for the greater. Specifications govern this distance by making it a function of the thickness of the plate or of the diameter of the rivet; one specification requires 2 diameters of the rivet for the least and 8 times the thickness of the plate for the maximum, but not to exceed 6 inches. As a usual thing, engineers desire the distance to be $1\frac{1}{2}$ inches for $\frac{7}{8}$-inch rivets, both limits being the same, and $1\frac{1}{4}$ inches for $\frac{3}{4}$-inch rivets.

The working point may or may not be on the plate. In many cases it is not. However, the distances must be measured from the working point. A familiar example of this is seen in the connection plates of the lateral systems of plate girders, see Fig. 54c.

In indicating the bevel, two methods are used. One is to state the actual rectangular dimensions, and the other is to reduce them so that the larger is 12 inches and the other a proportional part. For example, in Fig. 55 are given four working points and the lines connecting them. The plates are shown in outline. The bevel may be represented by the full dimensions or by taking the longest side as 12 inches and the other as 9.41 inches, or $9\frac{7}{16}$ inches, since it is unnecessary to get the bevel closer than the nearest sixteenth. The value 9.41 is computed as follows:

$$\frac{Y}{12''} = \frac{6'\text{-}6''}{8'\text{-}3\frac{1}{2}''}$$
$$Y = 9.41''$$

The method of indicating the bevel in feet and inches is much used, but indicating it in inches is preferable, since it is suitable to bench work. With the foot-and-inch method the floor of the templet shop has to be used in order to lay it out. The smaller values are advised for all work which can be worked on a bench.

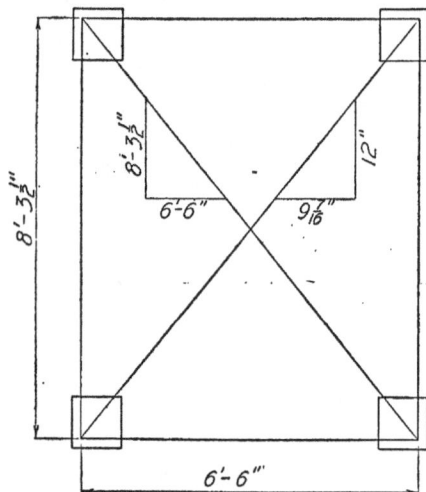

Fig. 55. Method of Indicating the Bevel

The number of rivets required in the connection plate in any direction must be sufficient to withstand the component of the main member attached to the plate. This can be easily determined by projecting the number of rivets in the diagonal against the line where the required number is to be placed. For example, let it be required to determine the number of rivets in the top of the plate in Fig. 56a, there being as shown 4 rivets in one diagonal and 3 in the other. Draw a diagram, as Fig. 56b, making o-1 and o-2 the same in slant as the diagonals above in 56a. Now with any scale whatsoever lay off four divisions from o towards 1 and three divisions from o towards 2 and project a line up from the last division mark to the horizontal line. Now measure o-a and o-b to the same scale which was used to lay off the divisions on the diagonal lines. There results 3.1 and 2.3 which means that 3.1 rivets are required for S_1 and 2.3 for S_2, or a total of 5.4 or 6 rivets for both.

It may be that the problem is as is shown in Fig. 56c. In this case the method of procedure is similar. Here, after drawing o-1 to the same slant as the member above, seven spaces to any scale are laid off and the projection made to the top and side. The results show that 6 rivets are required at the top and 3.2, or 4, are required at the side.

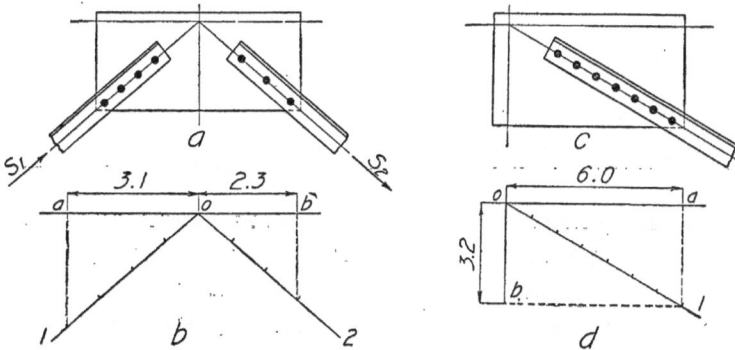

Fig. 56. Methods of Determining Number of Rivets in Connection Plate from the Diagonal

Other problems may be solved in a similar manner. The rivet spacing in the sides and tops is so arranged as to be equal and to fill out the plate, allowing the required edge distance. The plates should be kept rectangular as far as possible.

In many cases, as in roof truss or wind bracing work, the computed number of rivets will be two or less. In such cases three rivets should be put in in order to have a satisfactory joint which will not loosen under vibrations which are liable to occur.

Detailing of Combinations of Structural Shapes. The general methods to be followed are the same as those which have been given together with those which are exemplified in the discussions which

TABLE XIV

Thickness of Lacing Bars

Single Lacing $\left(t=\frac{c}{40}, \phi=30°\right)$		Double Lacing $\left(t=\frac{c}{60}, \phi=45°\right)$	
t	c	t	c
$\frac{1}{4}''$	$0'-10''$	$\frac{1}{4}''$	$1'-3''$
$\frac{5}{16}''$	$1'-0\frac{1}{2}''$	$\frac{5}{16}''$	$1'-6\frac{3}{4}''$
$\frac{3}{8}''$	$1'-3''$	$\frac{3}{8}''$	$1'-10\frac{1}{2}''$
$\frac{7}{16}''$	$1'-5\frac{1}{2}''$	$\frac{7}{16}''$	$2'-2\frac{1}{4}''$
$\frac{1}{2}''$	$1'-8''$	$\frac{1}{2}''$	$2'-6''$
$\frac{9}{16}''$	$1'-10\frac{1}{2}''$	$\frac{9}{16}''$	$2'-9\frac{3}{4}''$
$\frac{5}{8}''$	$2'-1''$	$\frac{5}{8}''$	$3'-1\frac{1}{2}''$

follow. In general, the combinations consist of plates or other shapes held together by angles, lacing bars, or tie plates, the size and section of the angles being determined in the design since they are part of the section of the member itself, while the lattice bars and tie or batten plates are chosen in accordance with the specifications employed. The specifications for lacing bars make their size a function of the distance between rivets. Table XIV gives the thickness of lacing bars for any distance between rivets.

Detailing of Beams. This is for the most part done on "Beam Sheets". These sheets are the size of the shop bills, 8½×14 inches, and have a printed heading and footing as on the shop bills. Between the heading and the footing are printed elevations and cross-sections of I-beams, as in Fig. 57, the number on a sheet varying with the number of dimension lines above and material below, i. e., from two to four. In some cases, those blank sketches are printed lengthwise of the sheet and then two only are placed upon a sheet. In case a channel is to be indicated, the draftsman blocks out one half of the section or end view, see Fig. 58, lower cut.

On these blank sketches the draftsman notes the rivets and rivet holes, puts on the connection and other angles, and shows all other information necessary for the complete fabrication of the beam ready for the structure of which it is a part. Figs. 58, 59, 60, and 61 are beam sheets which have been filled in, and illustrate very nicely the general principles.

The general rules regarding beam sketches are given in the following:

In all possible cases the holes in the end connections to the webs should be according to the standards given in the handbooks. If the

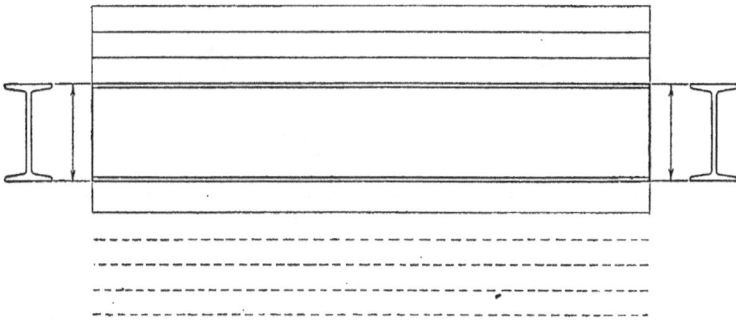

Fig. 57. Method of Detailing an I-Beam on Beam Sheets

connection is standard for that beam, no mention need be made of the fact, and if it is in the center of the web, no dimension is required, see Fig. 58, first view, left end.

If the connection is not in the middle of the web, but it is standard, the location of its center from the bottom should be given, see Fig. 58, first view, right end.

. If the connection is not standard, it must be noted and detailed as in Fig. 59, second view, and if it were not in the center of the web, its distance from the bottom should be given as in the case of the standard connection. In case there are holes in the outstanding leg, they should be shown as in Fig. 62. Where the leg against the web is standard and the outstanding leg is of the same punching, no dimensions need be shown, but the outstanding leg must be shown and the material notation of the angle put on as in Fig. 59, first view, right end.

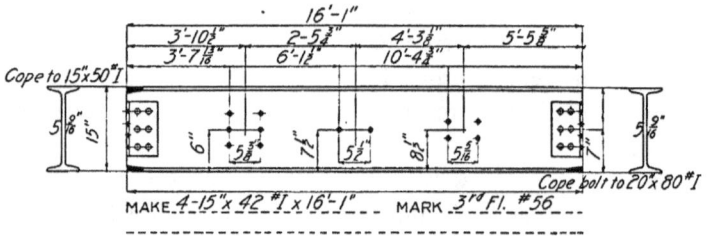

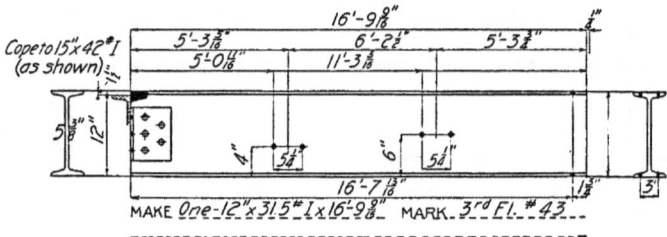

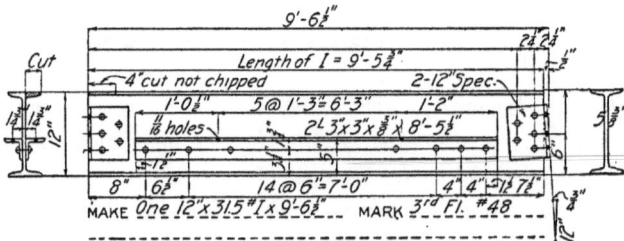

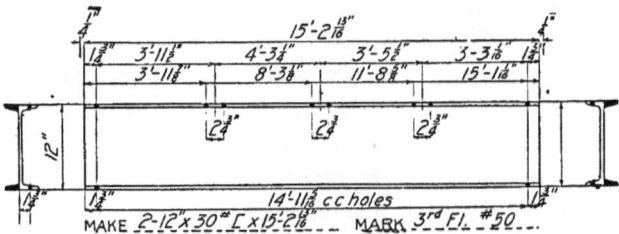

Fig. 58. Typical Beam Sheet Showing Dimensions Filled in According to Specifications

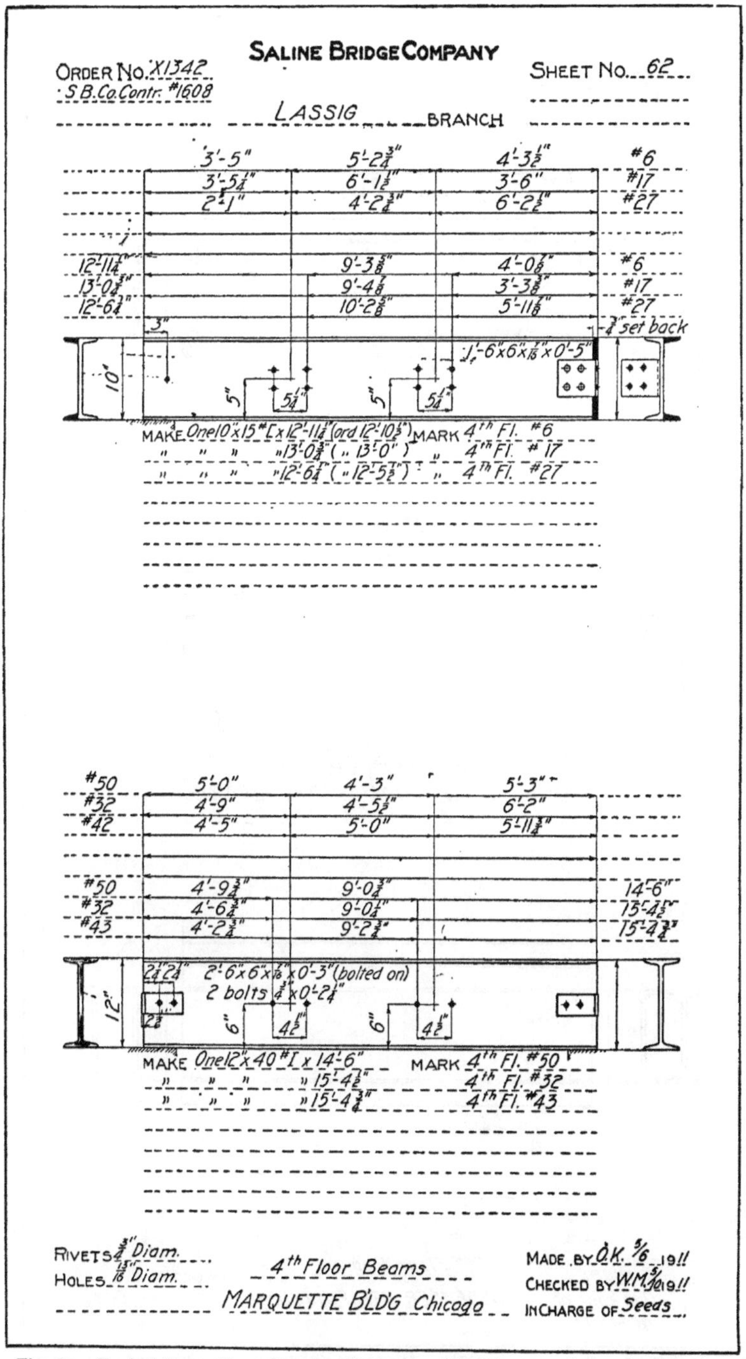
SALINE BRIDGE COMPANY
ORDER NO. X1342
S.B.Co.Contr. #1608
SHEET NO. 62
LASSIG BRANCH
MAKE One 10"x15# [x 12'-11¼" (ord 12'-10½") MARK 4th Fl. #6
" " " "13'-0¾" (" 13'-0") " 4th Fl. #17
" " " "12'-6¼" (" 12'-5½") " 4th Fl. #27
MAKE One 12"x40# I x 14'-6" MARK 4th Fl. #50
" " " " 15'-4½" 4th Fl. #32
" " " " 15'-4¾" 4th Fl. #43
2'-6"x6"x⅞"x0'-3" (bolted on)
2 bolts ¾"x0'-2¼"
RIVETS ¾" Diam.
HOLES 13/16 Diam.
4th Floor Beams
MARQUETTE B'LD'G Chicago
MADE BY O.K. 5/6 1911
CHECKED BY W.M. 5/10 1911
IN CHARGE OF Seeds

Fig. 59. Typical Beam Sheet Showing Dimensions Filled in According to Specifications

ORDER No. _X1592_
S.B.Co.Contr. #1806

_____ _SHIFFLER_ _____ BRANCH

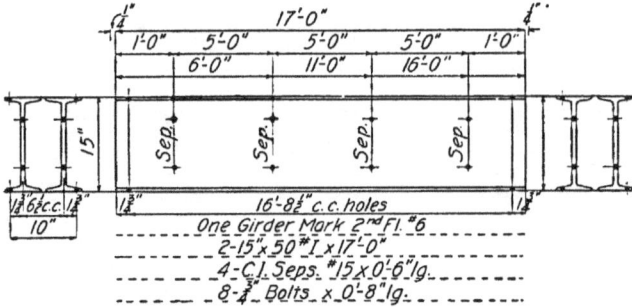

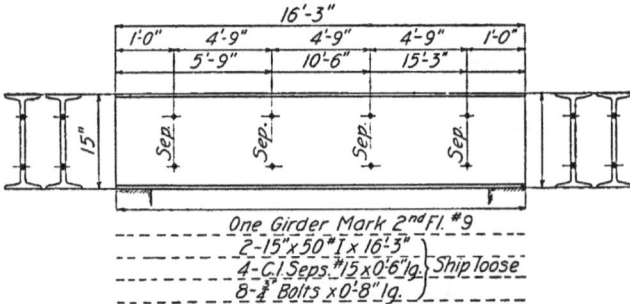

SHEET No. _32_

17'-0"

| d | 1'-0" | 5'-0" | 5'-0" | 5'-0" | 1'-0" | d |

6'-0"   11'-0"   16'-0"

15"

Sep.   Sep.   Sep.   Sep.

1½ 6 c.c.1½   1¾"   16'-8½" c.c. holes   1¾
10"

_____ One Girder Mark 2ⁿᵈ Fl. #6 _____
_____ 2-15"x50# I x 17'-0" _____
_____ 4-C.I.Seps. #15 x 0'-6" lg. _____
_____ 8-¾ Bolts x 0'-8" lg. _____

16'-3"

| 1'-0" | 4'-9" | 4'-9" | 4'-9" | 1'-0" |

5'-9"   10'-6"   15'-3"

15"

Sep.   Sep.   Sep.   Sep.

_____ One Girder Mark 2ⁿᵈ Fl. #9 _____
_____ 2-15"x50# I x 16'-3" _____
_____ 4-C.I.Seps. #15 x 0'-6" lg. } Ship loose
_____ 8-¾ Bolts x 0'-8" lg. _____

HOLES _¾"_ Diam.
PAINT _One coat of_
graphite

2ⁿᵈ Floor Beams

PASSENGER STA. Pittsburgh

MADE BY _A.B._ 2/2/19 11
CHECKED BY _X.Y._ 2/2/19 11
IN CHARGE OF _Doe._

Fig. 60. Typical Beam Sheet Showing Dimensions Filled in According to Specifications

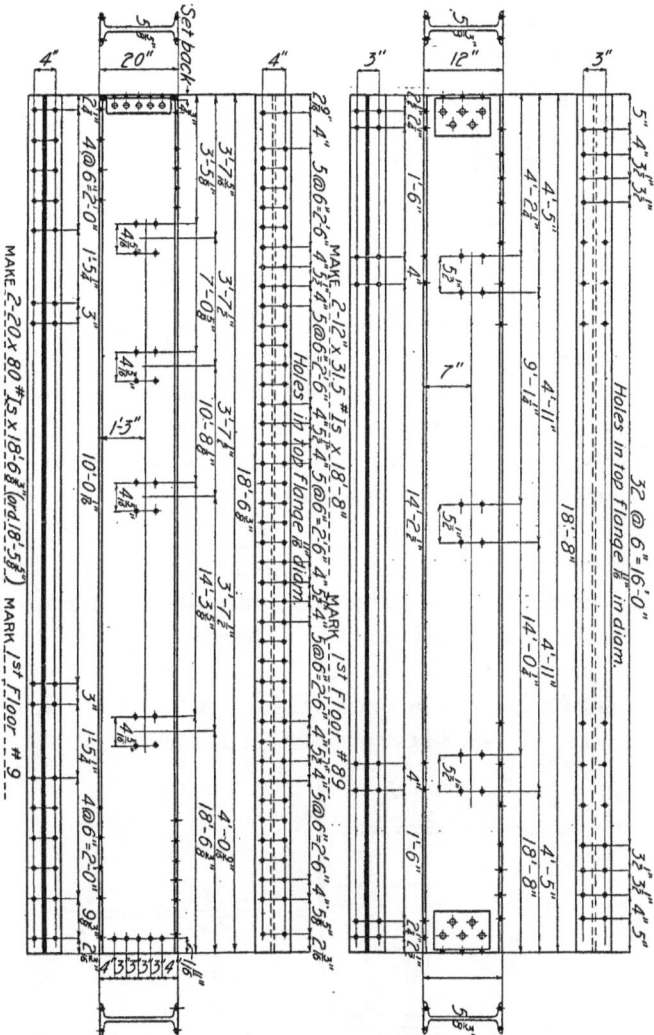

Fig. 61. Typical Beam Sheet Showing Dimensions Filled in According to Specifications

When beams are on a slight bevel, it is desirable to have the bevel taken up in the connection angles and the holes in the web of the beam

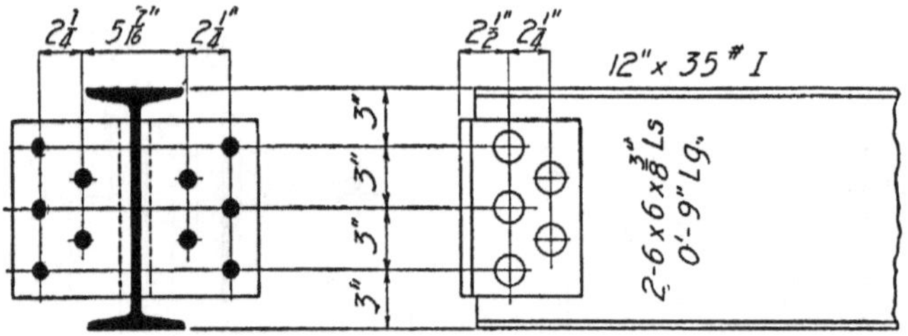

Fig. 62. Detailing Connection Plate When There Are Holes in the
Outstanding Leg

at right angles to the center line. The bevel should be indicated, see third view, Fig. 58, right end.

In case field connections to the web are made, as in cases where other beams are riveted to it, it is unnecessary to give the vertical spacing of the holes if the connection is standard. The horizontal distances and their number will designate which connection is required.

For example, in the first view, Fig. 58, the six holes 5⅜-inch centers show this to be a standard for 12-inch beams, while the four holes 5 5/16-inch centers indicate the standard connection for a 7-inch, 8-inch, 9-inch, or 10-inch beam. In all cases the vertical spacing will be 2½ inches. It should be noted that in all cases of standard connections of 8 holes or less in a vertical row the rivet spacing is 2½ inches, while all over 8 have a spacing of 3 inches.

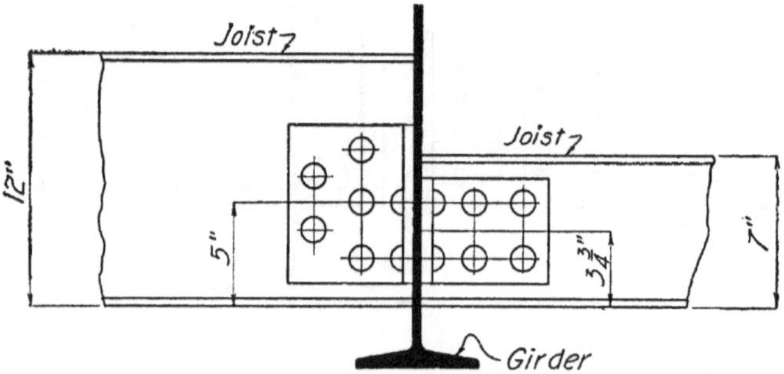

Fig. 63. Method of Bringing Beams to the Same Level on Main Girders

The centers of all groups of field holes above the bottom of the beam should be given.

Tie rods are put in in case no beams are riveted to the webs, to keep the beams from lateral motion. The holes for these are 4½ inches apart, and they are referenced as in Fig. 59, second view.

Where two beams are placed close together, they should be connected by "separators" to prevent lateral motion. When such is the case the holes are indicated as shown in Fig. 60. The various kinds of wall anchors are shown in the handbooks and in Fig. 20. Care should

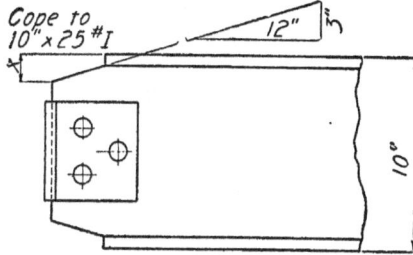

Fig. 64. Method of Coping a Beam Top and Bottom

be taken to provide for their connection to the beams when required.

When beams are used in building work, it is usually required that either the upper or the lower flanges of part or all of the beams be at the same elevation. When the girder or main beams are deep enough, the

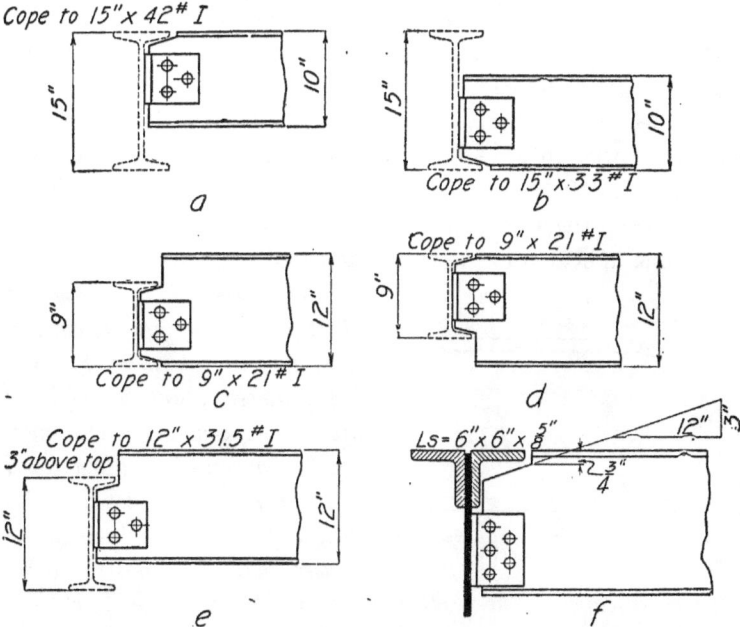

Fig. 65. Methods of Coping Beams to Fit Beams of Various Heights

joist top or bottom flanges may be brought to the same elevation as shown in Fig. 63 which shows a 12-inch and a 7-inch beam. The connection angles are in all cases arranged so that the rivets through the girder web and the smaller connection angles go through the connection of the larger beam also.

In case it is desirable to have beams so as to have all their tops or bottoms at the same elevation, it may be accomplished by an

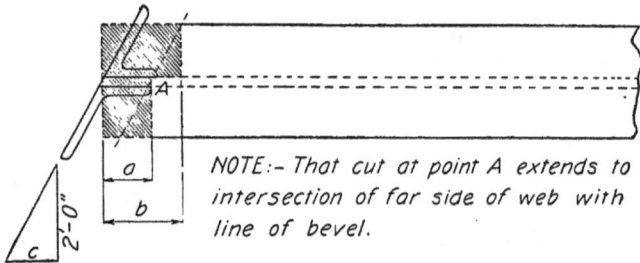

NOTE:- *That cut at point A extends to intersection of far side of web with line of bevel.*

Fig. 66.   Method of Cutting Flanges When a Beam is Coped on a Bevel

operation known as "coping" the beam. By coping is meant that the flange is cut back for a certain distance depending on the size of the beam which is to join the beam under consideration and the web is then cut down a distance X and sloped back on a bevel of 3 inches in 12 inches, see Fig. 64.

Fig. 64 shows a beam coped top and bottom to fit into another beam of its own depth. A beam may be coped on top only, Fig. 65a, or on bottom only, Fig. 65b. Other conditions of coping are shown in Fig. 65c—f, together with the ways of indicating them. Fig. 58 shows some indicated in the beam sketches.

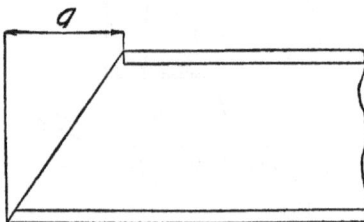

Fig. 67.   Method of Cutting an I-Beam or Channel to a Bevel

When a beam is to be coped on a bevel, the flanges are not cut to a bevel, but are cut as in Fig. 66. The distances a and b should be given allowing a $\frac{1}{4}$-inch clearance, and the portion of the beam coped is to be shown cross-hatched. This method of cutting to a bevel should be used whenever possible, whether the beam is coped to fit another or is simply cut to a bevel.

When a single beam or a girder formed of two beams having a cover plate riveted thereto is cut to a bevel, the cover plate should be sheared to the line of bevel and the beam should be cut as shown in Fig. 66.

When an I-beam of a channel is cut to a bevel across the depth, the cut should be made as shown in Fig. 67, and the distance "a" should be given.

Detailing of Roof Trusses. The first thing to determine in this respect is the outline of the outer line of the roof and the end, and the center depths. The chords should now be located by center lines corresponding to the gauge lines of the angles, or the center of gravity lines of the pieces, as the case may be. The above mentioned determinations may be obtained from the architect's drawing and from the stress sheet; and in many, if not most all cases, the center lines of the chords are shown on the stress sheet. The stress sheet may be an outline with the stresses and the sections on it, or it may and in fact should be as shown on Plate I. Here the designer, who is an experienced man, has shown the general details. It now remains for the draftsman to draw this up so that the shopmen can make it. After he has finished, the results will be as shown on Plates II and III, which will now be discussed in detail.

After the center lines of the chords are drawn in, the angles themselves should be drawn on by laying of the gauge lines on one side and then the other edge of the leg on the other side of the gauge line. After this the top chord should be divided into a certain number of equal parts at each of which a purlin is to be placed. This done, lines from these points should be drawn perpendicular to the top chord and their points of intersection with the bottom chord should be noted. From the intersection of the center one with the bottom chord to the apex or top, a line is now drawn, and this is the center line of the main interior tie, or tension member. The member itself should now be drawn on this gauge line. After this the other members should be drawn in as shown.

In order to proceed, the distances between the various points of intersection must be carefully computed, thus giving the remaining data necessary to compute the bevels, which should now be done.

In order to determine the length of the members and the sizes of the plates, it is now necessary to take each point of intersection where any members meet at any other than a right angle and make a layout of that joint to some large scale, say $1\frac{1}{2}$ to 2 inches to the foot. The customary $\frac{1}{4}$-inch clearance should be allowed where there is any liability of pieces touching and, after the ends of the

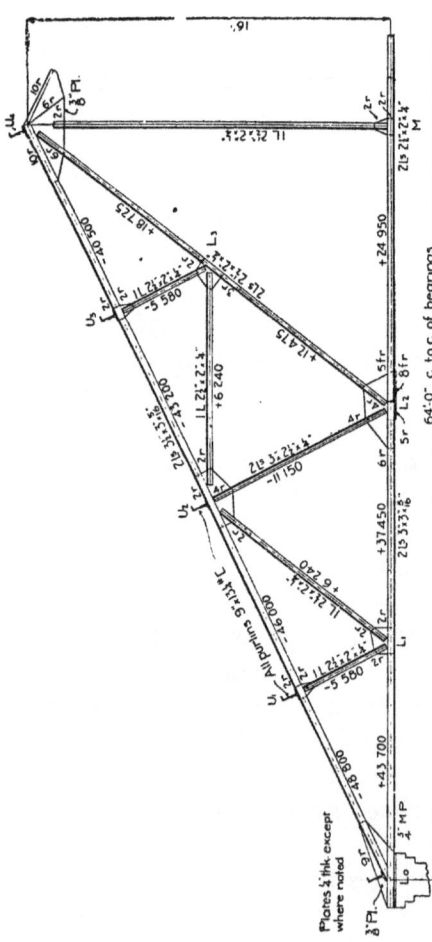

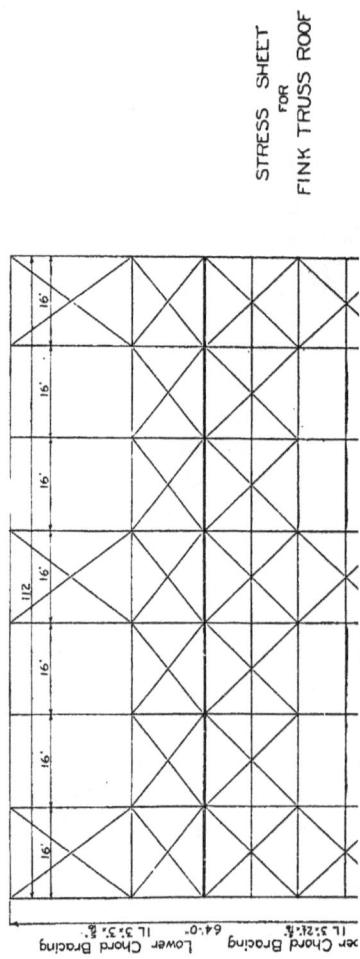

STRESS SHEET
FOR
FINK TRUSS ROOF

Make 2 Trusses as shown Mark T1
" 2 " " & noted, Mark T2R
" 3 " " to pair with T2R " T2L.
" 2 " as noted Mark T3
Make other half of truss like half shown with the
exception of Plate at Peak Mark as above
with subscript 'x' i.e. T1x, etc

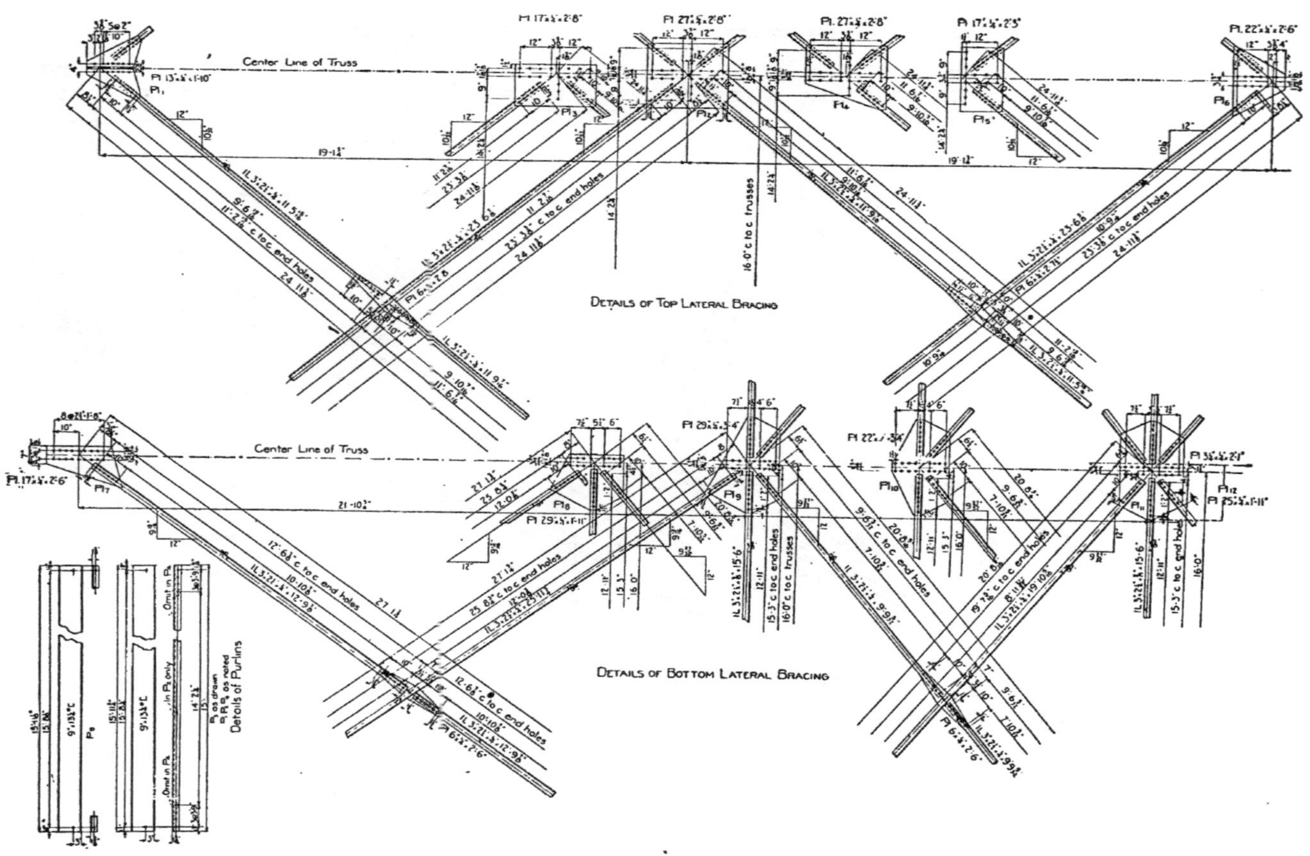

Plate IV. Work-Shop Drawing Showing Detail of Top and Bottom Lateral Bracing of a Roof Truss

various angles are drawn in, the first rivet is set back $1\frac{1}{8}$, $1\frac{1}{4}$, or $1\frac{1}{2}$ inches as the sizes of the angle and of the rivet allow, and the other spacing is so arranged as to make the size and shape of the plate advantageous and economical. The distance from the first rivet to the intersection is measured off and noted. After the layout for each joint has been made and the necessary dimensions of the plates and the distance from each intersection to the first rivet has been determined, the length of each member may be computed. This is equal to the length, intersection to intersection, plus the sum of the distances from the first rivet at the ends to the end of the member, minus the sum of the distances from the first rivet to the nearest, intersection. For example, in the main interior tie U_4 L_2, Plate I, the length, intersection to intersection, is $21'\text{-}10\frac{1}{2}''$, the distance from each first rivet to the end of the angle is $1\frac{1}{2}$ inches, and the sum of the distances from each first rivet to the nearest intersection is $(4\frac{1}{2}+9)$ $= 13\frac{1}{2}$ inches, which is $1'\text{-}1\frac{1}{2}''$. The length of the member is:

$$21'\text{-}10\frac{1}{2}''+2\ (1\frac{1}{2}'')-1'\text{-}1\frac{1}{2}''=21'\text{-}0''$$

At the point L_2, Plate I, a field connection must be made as well as at U_4 on account of the fact that the truss must be shipped in part in case the span is larger than 30 feet, the length of an ordinary gondola freight car. At L_2 both legs of the angle should be connected, the horizontal leg connection being by a plate. In case of riveted lateral bracing such as is used here, the connection plate may also be used as a splice plate, see Pls. 8, 9, and 10 in Plate III.

At point U_4 as many shop rivets are put in as there are field rivets required. This will keep the plate symmetrical, and will allow the same templets to be used for the top chord and main interior tie on both sides of the truss. This more than overbalances the cost of driving the few additional shop rivets.

At L_0 in this case the truss has been designed so that the rivets are symmetrical about the point of intersection and, therefore, only a sufficient number are required to take up the direct stress in the top and bottom chords. In many cases the end of a roof truss is as shown in Fig. 68, in which case the number of rivets $L_0\,L_2$ may be calculated from the equation:

$$n^2v - Rn = \frac{6\,R\,e}{p}$$

in which n=number of rivets required; v=allowable stress on one rivet; R=the vertical reaction; p=the rivet spacing in inches; and e=distance shown in Fig. 68.

The number of rivets in $L_0 U_1$ may be determined from the equation:

$$n^2 v - Sn = \frac{6 S e_1}{p}$$

in which S is the stress in $L_0 U_1$, and e_1 the distance shown in Fig. 68. These formulas allow for the stress due to eccentricity. The rivet spacing p is usually taken as 3 inches, although it may be taken as any value permissible by the specifications.

In the detailing of the lateral systems, Plate III, the same method of procedure as above mentioned should be followed. Care should

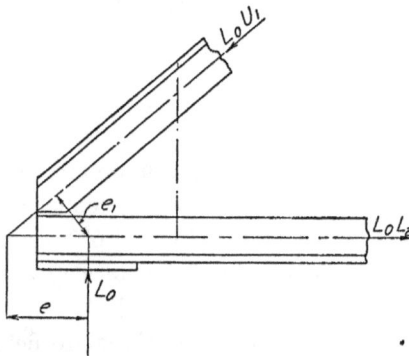

Fig. 68. Typical Detail for the End of a
Roof Truss

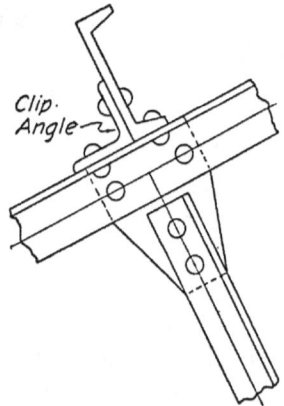

Fig. 69. Method of Riveting Clip
Angles for Carrying Purlins

be exercised in making the layouts for the lateral plates so that sufficient clearances are allowed, both in regards to clearances between members and clearances in rivet driving.

The purlins, or rafters, may be detailed directly upon the main sheet with the bracing or truss, or upon a beam sheet, preferably the latter. In Plate III they are upon the lateral sheet. These purlins should be riveted, not bolted to the chords of the trusses. In order to facilitate erection, clip angles should be riveted to the top chord as shown in Fig. 69 so that the purlin may be put in place and riveted up without having to hold it in place with ropes or chains.

Also by this method the purlin may be put in place and used as support for erection apparatus. In Plate III, the additional pair of holes at panel points of the top chord are for these clip angles.

After the draftsman has finished his drawing he should carefully check up all dimensions and bevels and inspect the drawing for errors in rivet clearances. The passing in of accurate detail drawings will soon result in a promotion to checker, a more pleasant position, but one with greater responsibilities attached.

Detailing of Plate Girder Spans. The information which the draftsman has to start with is in the form of the stress sheet. This may be as Plate V which is the latest and most approved form, or it will be like Plate VI. In both cases the number of rivets for the lateral connections are given, but on Plate V the rivet curve for the spacing in the flanges is given and also the curve of the total and dead load shears and moments.

As soon as a plate-girder stress sheet is turned over to the draftsman, he should lay it out at once and determine the exact location of the web splices, the stiffeners, and the cover plates and their lengths (if not given), should decide upon the lengths of the panels of the lateral bracing, and should also make layouts of the lateral plates, if possible, so that the material can be ordered at once if necessary. In making the above layout the following should be observed:

(1) Be careful in locating splices to see that they come at a panel point of the lateral system.

(2) Locate all splices and stiffeners with a view of keeping the rivet spacing as regular as possible.

(3) Have the panels of the lateral systems equal if possible. If not, have a smaller one at the ends of the girder, the remainder being of equal length.

(4) Stiffeners to which cross-frames are attached should have fillers.

(5) The outstanding leg of stiffener angles should have a gauge of $2\frac{3}{4}$ inches or more. This will enable the cross-frames or floor beams to be swung in during erection without spreading the girders.

(6) It is always best to use as few sizes as possible for stiffeners, connection plates, etc., and avoid all unnecessary cutting of plates and angles.

(7) Locate the end holes for laterals and diagonals so that they can be sheared by a single operation, see Fig. 70. This will, as a rule, throw the end rivet further back from the working point, and may increase the size of the connection plate, but it is desirable.

(8) It is preferable to have an even number of panels in the lateral system since the girders can in most cases then be made symmetrical or nearly so about the center.

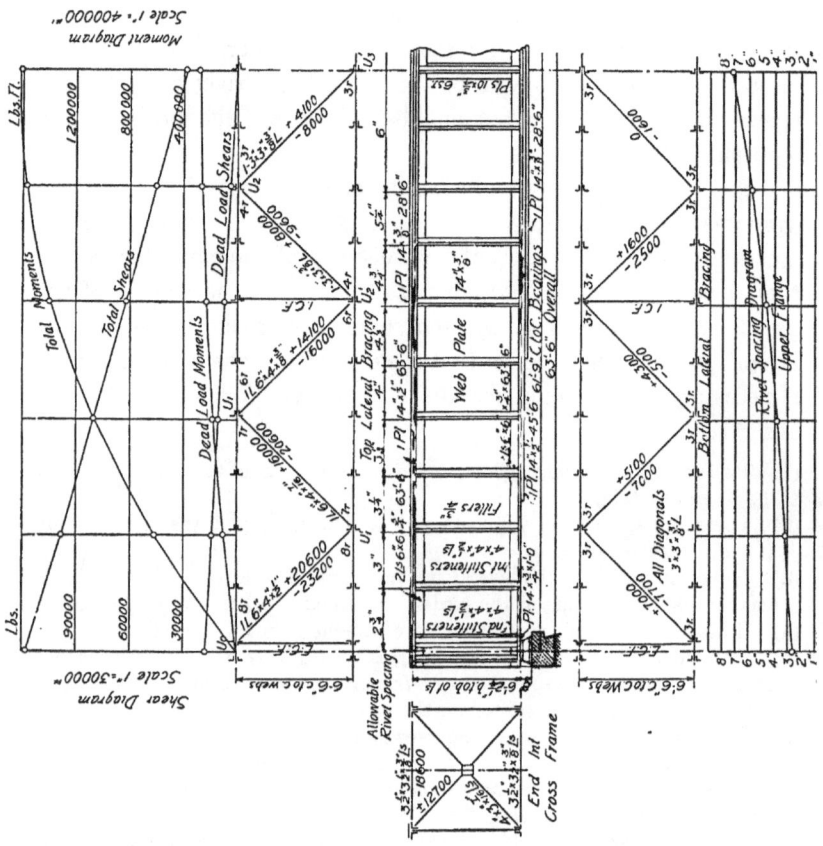

Dead Load W. L(124+10L)=4+200+Track
Live Load = E 40   =72300"
Specifications Cooper's 1901
All Rivets $\frac{7}{8}$" Diameter

Maximum Shears
Dead Load = 17800
Live Load = 100000
Total   = 117800

$\frac{117800}{9000}$ = 13.09 □" Net area required
$\frac{748\frac{1}{2}}{}$ = 2776 □" Gross area of web
$\frac{117800}{4920}$ = 25 Rivets

Maximum Moment in Lb Ft.
Dead Load = 375000
Live Load = 1340000
Total   = 1615000

$\frac{215000\cdot12}{72554}$ = 45400 D.L.Flange Stress
$\frac{1340000\cdot12}{72.554}$ = 221500 L.L.Flange Stress
$\frac{45400}{20000}$ = 2.27 □"
$\frac{221500}{10000}$ = 11.20 □"
2447 = Net area req in flange

2 Ls 6"×6"×$\frac{3}{4}$" = 13.88 □" Net
1 C.Pl 14×$\frac{1}{2}$" = 6.00 □" Net
1 C.Pl 14×$\frac{3}{8}$" = 4.50 □" Net
Total = 24.38 □" Net

Moment Diagram
Scale 1"= 400000"

Shear Diagram
Scale 1"= 30000"

Lbs Ft.
1200000
800000
400000

Total Moments
Total Stress
Dead Load Moments
Dead Load Shears

Lbs
90000
60000
30000

Web Plate 74×$\frac{3}{8}$"
Pls 10×$\frac{3}{8}$" 63$\frac{1}{2}$"
Fillers $\frac{5}{16}$"
Int Stiffeners 4×4×$\frac{3}{8}$"
End Stiffeners 4×$\frac{3}{4}$×$\frac{1}{2}$"

Top Lateral Bracing
Bottom Lateral Bracing
All Diagonals
Upper Flange
Rivet Spacing Diagram

Allowable Rivet Spacing
6-6 C to C webs
6-6 C to C webs

End Int Cross Frame

L V R R BRIDGE No.358A
SINGLE TRK. DECK GIRDER
Over Raritan R Blake Co, N.J.
61'-9" C to 6.Bearings
CAPITOL BRIDGE CO.
Pekin Plant

Cont No.---   In Charge of.-------
Made by----.   Date----.-------
Checked by----.Date----.-------
Order No. ---   Sheet

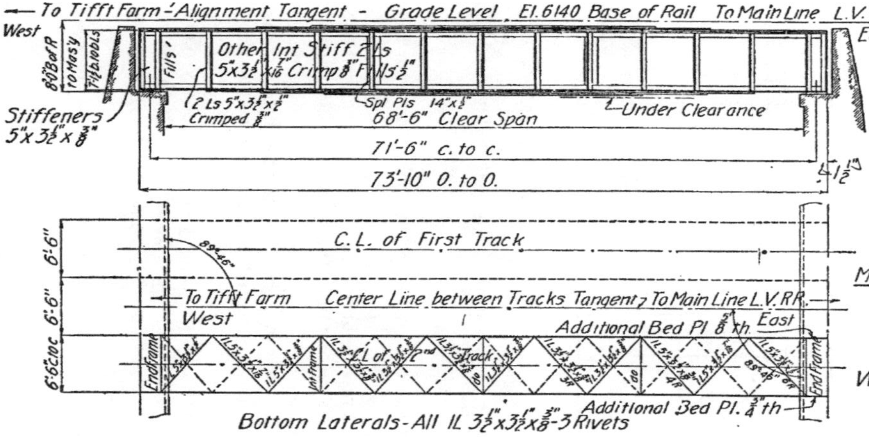
To Tifft Farm - Alignment Tangent - Grade Level El. 6140 Base of Rail To Main Line L.V.R.R.
West
East
8'0" B of R
10 Mas'y
Other Int Stiff ∠s 5"x3½"x7/16" Crimp 3/8" Fills ½"
2 Ls 5"x3½"x½" Crimped 3/8"
Spl Pls 14"x½"
Stiffeners 5"x3½"x3/8"
Under Clearance
68'-6" Clear Span
71'-6" c. to c.
73'-10" O. to O.
C.L. of First Track
89°46'
To Tifft Farm
West
Center Line between Tracks Tangent To Main Line L.V.R.R.
Additional Bed Pl 5/8" th.
East
End frame
Int frame
Additional Bed Pl. 3/4" th
Bottom Laterals - All IL 3½"x3½"x3/8" - 3 Rivets
6'6"
6'6"
6'6" c to c

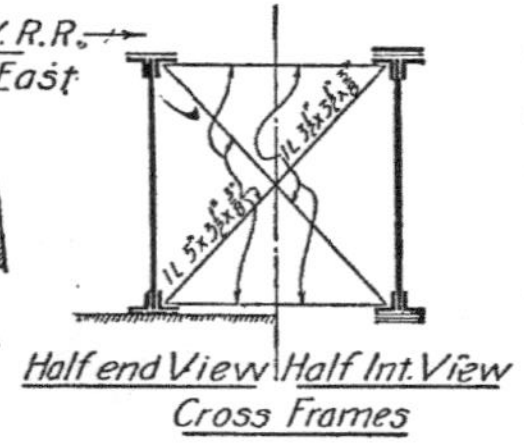
Half end View
Half Int. View
Cross Frames

Specifications: L.V.R.R. 1906

Material:- Soft O.H. Steel.

Rivets:- 7/8" φ

Ties:- 8"x9" notched to 8½"

See Purchasers Plans R.B.S. 893 & R.B.S. 873 Also B.R.A. 789

MAIN GIRDER

Max. Shear		Max. Mom.	
D = 26300#		D = 5623000"#	
L = 140500		L = 26565000	
166800#		L + D/2 = 29377000"#	
@ 5500 = 30.3 gr.		÷ 848 = 346000	
Web 85 x 3/8 = 31.3 gr.		@ 9000 = 38.4	

2 Ls 6"x6"x7/8" = 19.5 gr. - 16.0 net

1 Cov. 16"x9/16" = 9.0 gr = 7.9 - Full Top - 58'-0" Bot.

1 " " = 9.0 = 7.9 - 48'-0" T. & B.

1 " 16"x½" = 8.0 = 7.0 - 33'-0" T. & B.

38.8

Assumed Dead Load

Steel	1070#
Track	400
	1470# per lin. ft.

Assumed Live Load Cooper's E 50

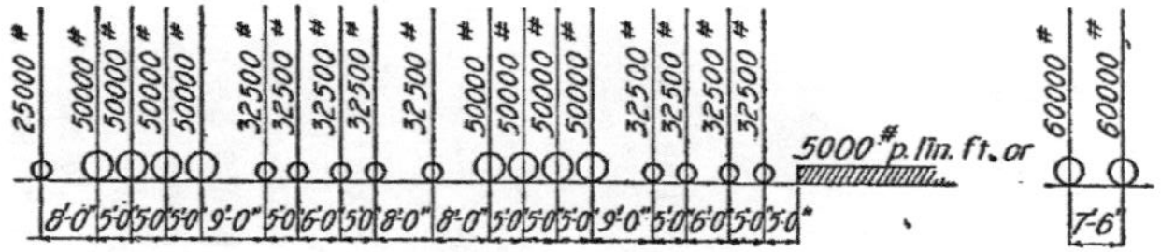
25000#
50000#
50000#
50000#
50000#
32500#
32500#
32500#
32500#
32500#
50000#
50000#
50000#
50000#
32500#
32500#
32500#
32500#
5000# p. lin. ft. or
60000#
60000#
8'0" 5'0" 5'0" 5'0" 9'0" 5'0" 6'0" 5'0" 8'0" 8'0" 5'0" 5'0" 5'0" 9'0" 5'0" 6'0" 5'0" 5'0"
7'6"

Approved as to Order C382 June 8-'06

Inquiry No. E 21874

L.V.R.R.

LEHIGH. & LAKE ERIE R.R.

BRIDGE LL. 444 B.

One D.T. Dk. Pl. Gdr. Span 73'-10" o. to o.

Stress Sheet

Scale 1/8" & 1/4" = 1'-0"

(9)   The rivet spacing curve should be constructed if it is not given on the stress sheet.

In addition to the above the following rules which apply directly to the detailing should be followed.   They are:

(1)   The second pair of stiffeners over the ends of the bed plate shall be so placed that the plate will extend not less than 1 inch beyond the outstanding leg.

(2)   If spans are on a grade, unless otherwise specified, put the bevel in the bed plate or masonry plate and not in the base or sole plate, sometimes called the bearing plate.

(3)   In short spans, 50 feet or less, put slotted holes for anchor bolts in both ends of the girder.   This will usually be covered by a clause in the specifications.

(4)   In square spans show only one-half, but give main dimensions such as "overall" and "center to center" and lengths of cover plates for the entire span.

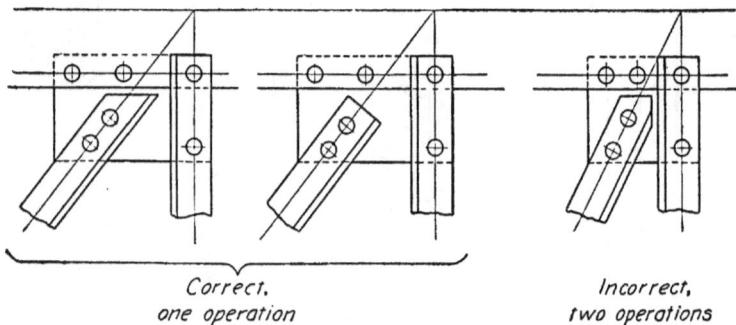

Correct,
one operation

Incorrect,
two operations

Fig. 70.   Method of Locating End Holes for Laterals and Diagonals so that They May be Sheared by a Single Operation

(5)   The girder detailed is always the far girder and is looked at from the inside.

(6)   If a span has no lower lateral bracing, only sufficient of the ends of the girder are to be shown in order that the detail of the base plate and its connection to the flange may be shown.

(7)   If the fillers become 12 to 15 inches wide, they become too heavy to be slipped in in the field and they should be riveted in place in the shop with at least two countersunk rivets.

(8)   When the ends of two girders meet on the same pier the masonry plate should be made continuous, that is, one plate to extend under both spans. Never make the base plates continuous since they could not be riveted up in the field.

(9)   Detail the bed (masonry) plate separately, never show it in connection with the base plate.

, At least two sheets are necessary to complete the detail drawings of any girder span, viz, (1) the Floor, Masonry, and Erection Plan, Plate VII, and (2) the Girders and Bracing, Plate VIII, although in many cases the information on this sheet is put on two sheets, the girder on one and the bracing on the other. The first sheet should show:

(1) A cross-section of the floor.
(2) A longitudinal view of the floor.
(3) A side elevation of the floor.
(4) The angle of skew and the width of the bridge seat.
(5) The elevation of the bridge seats and the grade of base of rail.
(6) The marking diagram.
(7) All clearances.
(8) Other essential information.

In the marking diagram all members which are entirely alike should be given the same mark. It may be, and usually is a fact, that all marks can not be put on the marking diagram until the detail drawing is done since then and only then is it possible, especially with the plates, to determine all members which are alike. Only those members which are shipped loose are given a mark. Thus it is seen that while each connection plate has a mark, only the entire cross-frames are given one mark since the members which compose them are all shop-riveted together. "Other essential information" is seldom required. In this special case there is shown another track which it is proposed will be put in in the future. Another case is where each end of the span has a different height from the base of rail to the masonry. In such cases this should be shown.

On this sheet should be shown the masonry plates, and if the ends are supported on cast-steel bases the height of these and also the dimensions of the base should be given.

The following general rules apply to the second sheet, Plate VIII.

(1) At the top of the sheet show a top view of the span with cross-frames, laterals, and their connections complete, the girders being placed at their proper distances apart.
(2) Below this show the elevation of the far girder from the inside, with all field holes in the flanges and stiffeners indicated and blackened in.
(3) If the span has lower lateral bracing, show below the elevation a horizontal section of the span just above the tops of the lower flange angles. On this drawing show the lower lateral bracing.
(4) Cross-frames shall, whenever possible, be detailed on the right hand of the sheet in line with the elevation. The frame shall be of such a depth as

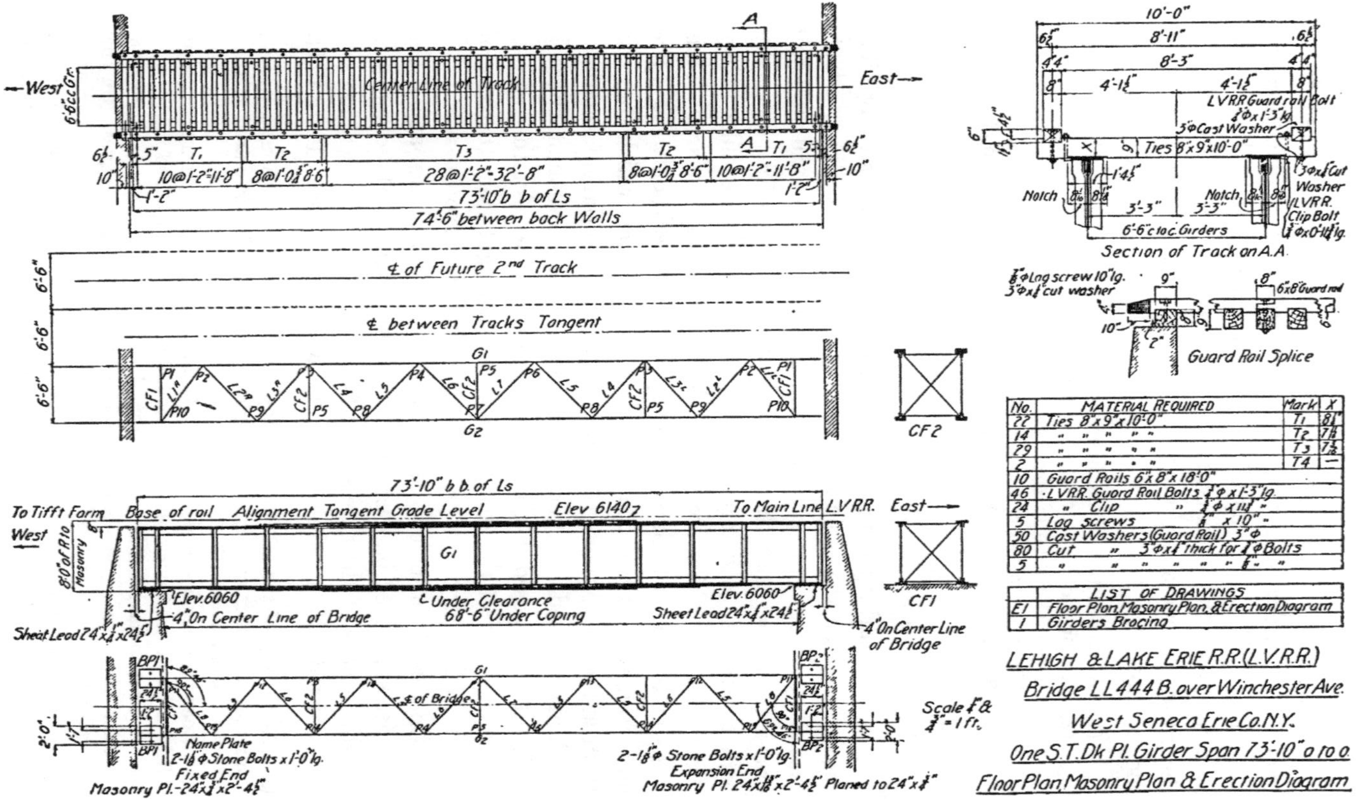

Plate VII. Floor, Masonry, and Erection Plan for Plate-Girder Span

to permit it being swung into place without interfering with the heads of the flange-rivets in the girders.

(5)   Always use a plate, not a washer, at the intersection of the diagonals of cross-frame.

The various parts of Sheet 2, Plate VIII, will now be taken up in detail and described and commented upon.

The Webs. As a usual thing webs are never specified in fractions of an inch. If so, the next inch in width must be ordered and then after the flange angles are riveted on, the projecting portion is cut off—an expensive operation. Webs are ordered in even inch widths and the distance back to back of angles is made $\frac{1}{4}$ inch or $\frac{1}{2}$ inch greater than the width of the web plate. This is sufficient to prevent any irregularities in rolling from projecting above the flange angles. Some engineers favor the web planed down from the greater width and claim that the bearing of the web on the sole or base plate thus obtained is a great advantage. The advantage is slight, however, and unless specially instructed to detail it that way, it should not be done. The web splices should, as before mentioned, be at a panel point of the lateral system. In some cases the web plates butt up against each other, being planed to an even bearing. In most cases, however, the ends of the webs are sheared off and the customary $\frac{1}{4}$-inch clearance is allowed. In this case the sum of the lengths of the webs as given is $2 \ (25'\text{-}\frac{3}{4}'') + 23'\text{-}5\frac{3}{4}'' = 75'\text{-}9\frac{1}{4}''$, while the "overall" distance is $73'\text{-}10''$ or $\frac{3}{4}$ inch less, which is taken up by the distance between webs at the splice and by the small amount, $\frac{1}{8}$ inch, which web is below the backs of the angles at the ends. This shows the webs to be $\frac{1}{4}$ inch apart at the splices. It is unnecessary to place any dimensions or notes on the drawing calling attention to this fact since the shop will make this allowance unless instructed otherwise. In case the webs are to be close together, a note must be placed on the drawing at the splice, reading "Webs planed to even bearing."

Web Splices. Web splices may be of two forms, viz, that as indicated on Plate VIII which takes shear only, and the moment web splice. The proper manner to detail a moment web splice is as shown in Fig. 71. In the simple shear splice both the splice stiffeners and the splice plates may and should have the same spacing as the intermediate stiffeners, and the rivet lines should be spaced so as to

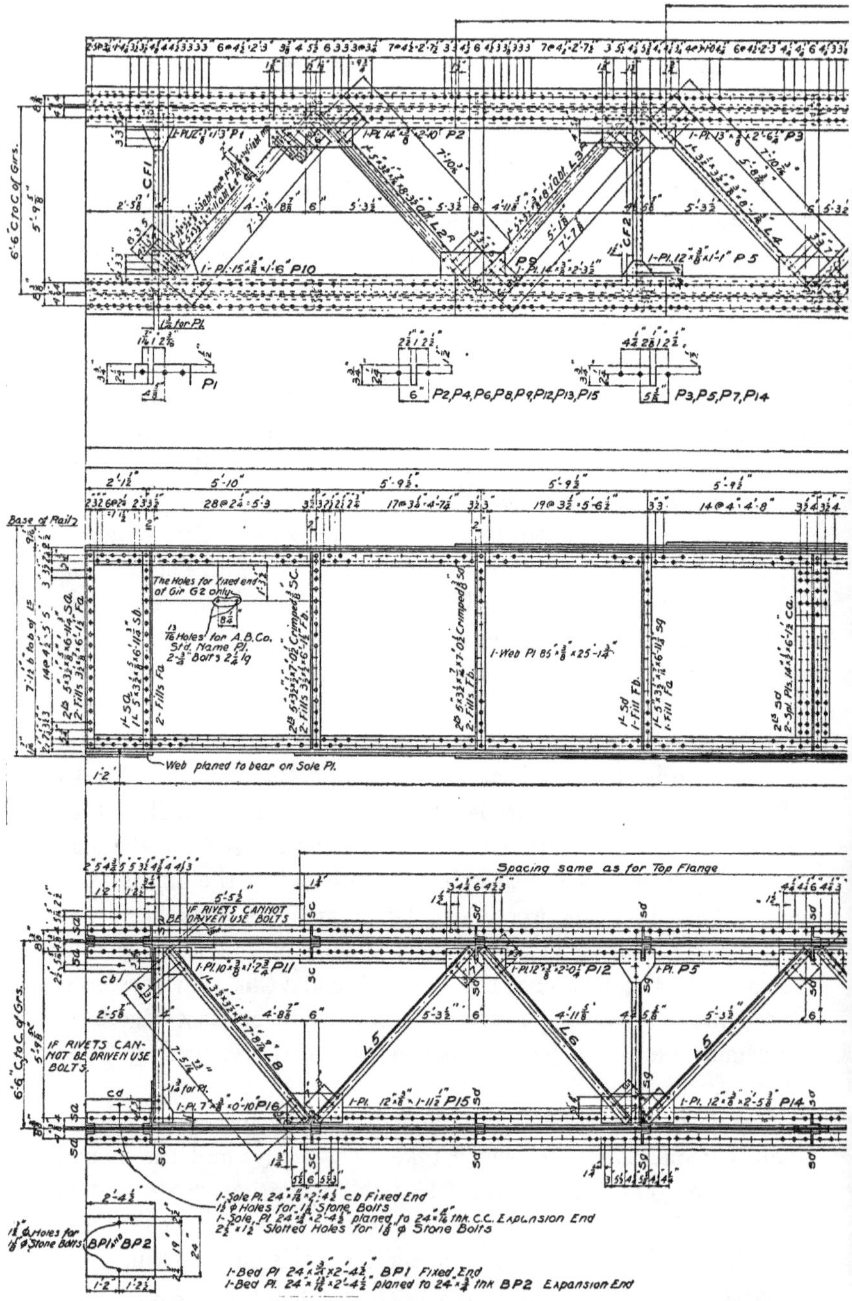

Plate VIII. Detail Drawing of Girders and Bracing for Plate-Girder Span'

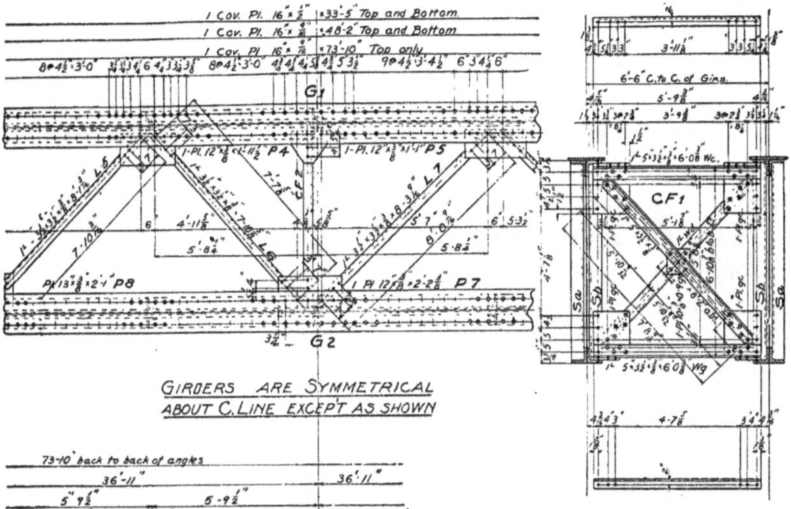

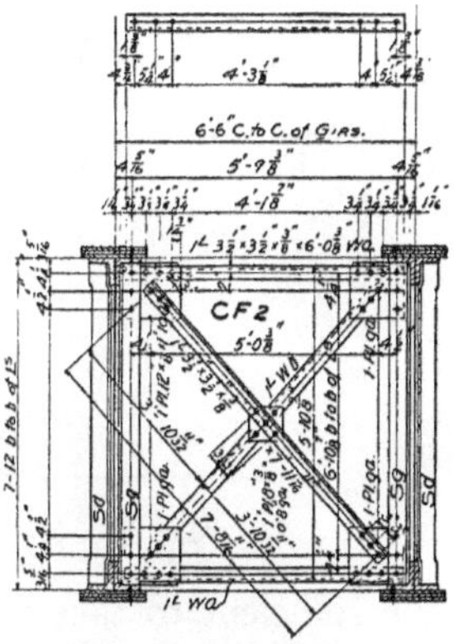

1 Cov. Pl. 16" x 1/2" x 33'-5" Top and Bottom.
1 Cov. Pl. 16" x 9/16" x 48'-2" Top and Bottom
1 Cov. Pl. 16" x 7/16" x 73'-10" Top only
G1
G2
P4
P5
P7
P8
L5
L6
L7
CF1
6'-6" C. to C. of Girs.
GIRDERS ARE SYMMETRICAL ABOUT C. LINE EXCEPT AS SHOWN

73'-10" back to back of angles
36'-11"
36'-11"
G1
2Ls 6"x6"x7/8"x73'-10"
71'-6" C. to C. of Bearings
1 Cover Pl. 16"x7/16"x59'-0" Bottom only
P4
P5
P7
L5
L6
L7

6'-6" C. to C. of Girs.
CF2

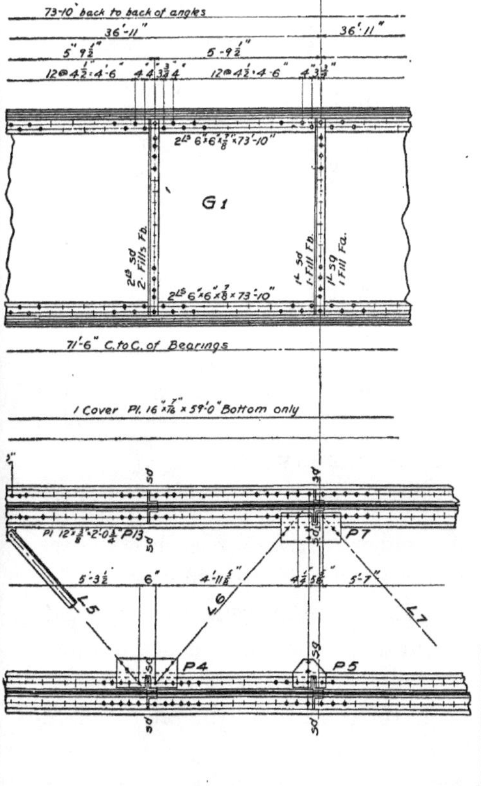

Material Soft O.H. Steel
Rivets 7/8" diam
Open Holes 5/16" diam, except as noted

LEHIGH AND LAKE ERIE R.R. L.V.R.R.
BRIDGE L.L.444B, OVER WINCHESTER AV
WEST SENECA, ERIE CO, N.Y
One S.T. Deck Pl. Gir. Span 73'-10" o to o
GIRDERS AND BRACING
Scale 1/2" = 1 Ft.

correspond to the spacing in the flange angles. In the moment splice this should be done if possible, but this is seldom the case. However, in case of more than one splice occurring in half of the girder, they should all be made alike, being figured for the one with the greater stress. Since a splice plate is a species of filler, it should be given a mark so that in case of other splices occurring the mark and not all the dimensions should be placed upon it.

Stiffeners. All stiffeners except the second from the end should have the outstanding leg on the side of the gauge line away from the center of the girder. As a rule, the end stiffeners should have enough rivets to take up the end shear, and the intermediate stiffeners should

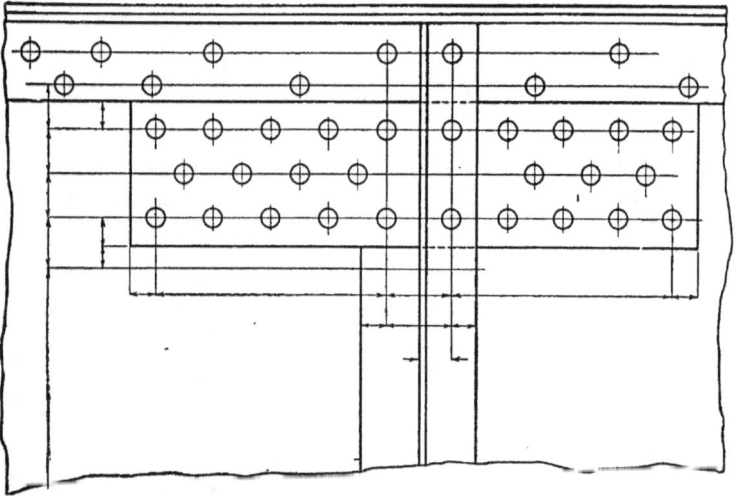

Fig. 71   Method of Detailing a Moment Web Splice

have sufficient to take up the shear at that point. This would, if carried out, require a different number of rivets in each stiffener. Common practice requires that the spacing in all stiffeners should be the same and that this spacing should be the same as in the end stiffeners. In some cases, such as in heavy girders, it is not possible to do this on account of the large number required in the end stiffeners, and the rivet spacing is then made the same in all the intermediate stiffeners.

The rivet spacing should not exceed $4\frac{3}{4}$ or 5 inches at the most and should be so placed that it should be symmetrical about the center. When the web plates are of even inch width, the $\frac{1}{4}$ inch may be put into one odd space at the center in order to avoid $\frac{1}{8}$ inch in the spacing. It may be necessary to put in a few more rivets than are computed as necessary, but the advantage gained by thus making the punching of the plate on the multiple punch possible, makes this advisable. Fig. 72 shows this method of detailing. In order to make the shear plate at the web splice efficient to some degree in withstanding the moment—for although it is not computed to take moment yet it does in reality —the rivets near the flange are placed close together for a few spaces. If the space changes after that, it should increase towards the middle of the web, except in such a case as Fig. 72 where the center space may or may not be as great as those on either side of it.

In double-gauge flange angles the rivet in the stiffener should be in the inner guage line of the flange angle as shown, and no rivet should come closer than $1\frac{1}{2}$ inches to the end of a filler.

Each like stiffener should be given a mark and in case others of the same kind both in size and punching occur, the mark may be used instead of the material notation and dimension. Those crimped will be given different

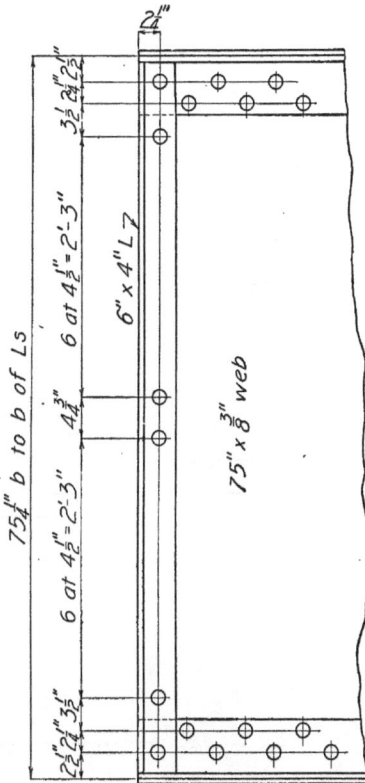

Fig. 72. Method of Detailing Web Plate Stiffeners

mark even if size and punching are the same. It should be noted that some of the stiffener angles differ only from the fact that they have holes in their outer leg to which the cross-frames are connected, hence a different mark.

The length of stiffeners listed on the drawing is the distance inside of flange angle legs. Without further instructions they will be ground by the shopmen so as to have a snug fit.

Fillers. Fillers are placed under angles that are crimped since the angles are only crimped $\frac{3}{8}$ inch and not the entire $\frac{7}{8}$ inch which is the thickness of the flange angles. The fillers are given marks for the same reasons and in accordance with the same rules that apply to stiffeners.

Flange Angles. In case of double gauge on the 6-inch flange angle it is better to put the 2½-inch gauge on the inside, no matter

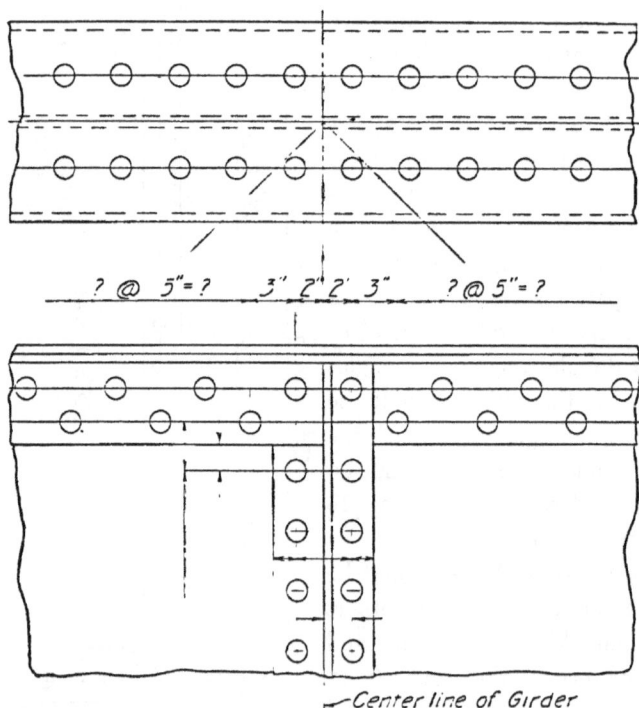

? @ 5"=?   3" 2" 2" 3"   ? @ 5"=?

⌐Center line of Girder

Fig. 73.   Method of Detailing Rivet Spacing with Flange Angles

what the thickness may be, since by this operation the rivets in the horizontal flange, providing that is a double-gauge line, may be more advantageously spaced on account of the fact that the required stagger will be less.

The rivet spacing in the vertical leg of the flange angles should increase from the end towards the center and should remain the same, as far as possible, between any two stiffeners, any changes necessary being made near the stiffeners. Since a rivet must always be in the inner gauge line at a stiffener, an even number of spaces must be between any two stiffener gauge lines, since the rivets must stagger. This brings one rivet in the center of the girder, which can not occur in case there is a splice at the center of the girder. The stagger may then be broken as in Fig. 73, the stiffener angle being placed as shown and the rivet spacing being symmetrical on each side of the center of the girder. .

Between the stiffeners at the end, the spacing should be the same as it is between the next two stiffeners.

The spacing at any point should never exceed the computed spacing unless constructive reasons require it. On account of rivet-driving clearances, a $\frac{7}{8}$-inch rivet can not be driven any closer than $1\frac{1}{4}$ inches to another member. Therefore, rivets can not be driven any closer to the stiffener than $1\frac{1}{4}$ inches, see Fig. 74. For this particular sized stiffener, the minimum spacings next to it will be $3\frac{1}{4}$ inches and $2\frac{3}{4}$ inches as seen in Fig. 74. The rivet-spacing multiplication table, Table X, will be found very helpful in spacing the rivets here.

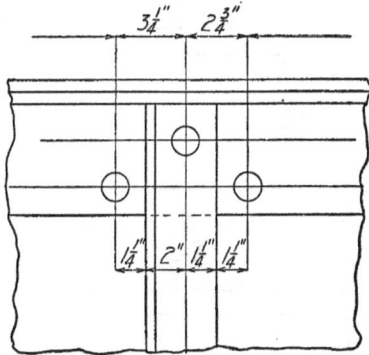

Fig. 74. Minimum Rivet Spacing for Stiffener Angles

Since the single gauge is used in the top flange and, according to the stress sheet, two rivet holes are taken out of each angle, it is possible to space the rivets in the outstanding leg and cover plates without reference to those in the other leg of the angle, due care being taken that they do not come closer than $1\frac{1}{4}$ inches to the out-standing stiffener leg. No special rule governs the spacing in the cover plates, the only requirement being those of the specifications, and that the number of rivets from the center of the span to the end of the cover plate or the number of rivets from the end of one cover plate to the end of another shall be

$$n = \frac{(\text{net area of cover plate}) \ s}{v}$$

where n = the number required; s = the allowable unit flange stress:

and r = the value of a rivet in single shear or bearing in the cover plate. whichever is the smaller. For the first cover plate on top of the flange angles this equation gives

$$n = \frac{[\,16 \times \frac{7}{16} - 2\,(\frac{7}{8} + \frac{1}{8})\,\frac{7}{16}] \times 10000}{6013}$$

$$= 13$$

which shows the number 78 to be amply sufficient in this respect. A clause in most specifications requiring the maximum spacing to be not greater than 16 times the thinnest plate and not greater than 6 inches. further governs the number. which would be 50 by this re-

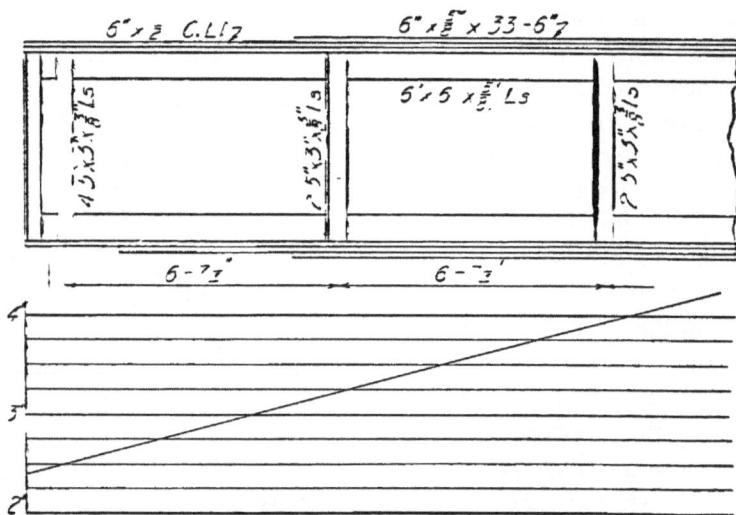

Fig. 75.   Typical Stiffener and Rivet Spacing Diagram

quirement. Most engineers, notwithstanding the specifications, require the majority of the spacing to be within 5 inches.

In case the spacing in the top flange is on a double-gauge line. care must be taken to see that the minimum stagger, Table IX, is not violated. In such cases it is customary to place a rivet in the inner gauge line of one leg opposite a rivet in the outer gauge line of the other leg. and to do this until a stiffener interrupts, when spacings are made with the observance of Table IX, until the rivets can be placed opposite again.

In order to illustrate the above principles in regard to spacing when double-gauge lines are used on both legs and the maximum spacing for any particular distance is shown by the rivet curve, an example will be given. Let the stiffeners and the rivet-spacing diagram be as in Fig. 75. This shows the allowable rivet spacing to be 2½ inches at the second, 3¼ inches at the third, and 3⅞ inches at the fourth stiffener, the distance between stiffeners being 6'-7¼''. Let it be required to determine the rivet spacing between the second and fourth stiffeners

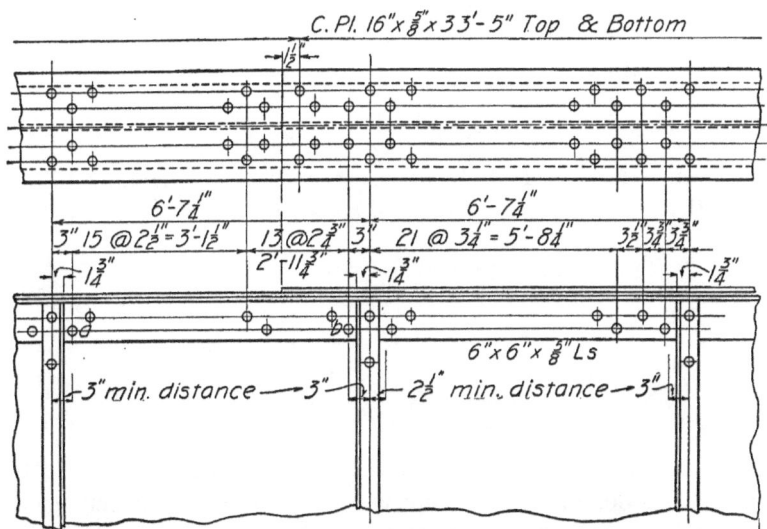

Fig. 76. Detail Drawing Showing Determination of Rivet Spacing between Second and Fourth Stiffeners

Since the stiffener angles have a 3-inch leg on the web, the gauge of which is 1¾ inches, and no rivet can be driven closer to the edge of the leg on the web or to the outstanding leg than 1¼ inches, no rivets can be driven closer to the gauge than 3 inches and 2½ inches on the sides of the outstanding leg and the edge of the other leg, respectively, see Fig. 76. Since 3 inches is the minimum distance it must be used at stiffener (2) notwithstanding the fact that the spacing diagram requires not less than 2½ inches. This leaves (6'-7¼'') −3''=6'-4¼'' from that rivet to the one in the gauge at the top of stiffener (3), no attention being paid to 3 inches, the minimum dis- tance here, since it is less than the 3¼ inches required by the diagram.

An odd number of spaces must be used since the last rivet is on the other gauge line; and from the rivet-spacing diagram it is seen that the spacing can not exceed $2\frac{3}{4}$ inches until half way between the two stiffeners, and that a space or two of $3\frac{1}{4}$ inches would be allowed at stiffener (3).

By consulting Table X it is seen that 29 spaces at $2\frac{1}{2}$ inches are equal to $6'$-$0\frac{1}{2}''$. Now $(6'$-$4\frac{1}{4}'') - (6'$-$0\frac{1}{2}'') = 3\frac{3}{4}''$ or 15 fourths ($\frac{15}{4}$), from which it is seen that if $\frac{1}{4}$ inch was added to 15 of the 29 @ $2\frac{1}{2}''$, the result would be all that is desired; but this would leave the last space $2\frac{3}{4}$ inches and by Fig. 76 it is seen that it must be at least 3 inches. By making the last space 3 inches, which is $\frac{1}{2}$ inch, or $\frac{2}{4}$ greater than $2\frac{1}{2}$ inches, there remain 28 spaces between rivet a and rivet b, Fig. 76, and only $\dfrac{15}{4} - \dfrac{2}{4} = \dfrac{13}{4}$ left. If, therefore, $\frac{1}{4}$ inch be added to 13 of the $2\frac{1}{2}$-inch spaces, making 13 of $2\frac{1}{2}'' + \frac{1}{4}'' = 3\frac{3}{4}''$ each, the spacing will be correct. It is:

$$
\begin{array}{llll}
1 \text{ space at} & 3'' & = 0'\text{-}3'' \\
15 \text{ ``} & \text{`` } 2\frac{1}{2}'' & = 3'\text{-}1\frac{1}{2}'' \\
13 \text{ ``} & \text{`` } 2\frac{3}{4}'' & = 2'\text{-}11\frac{3}{4}'' \\
1 \text{ ``} & \text{`` } 3'' & = 0'\text{-}3'' \\
\hline
& & \text{Total} = 6'\text{-}7\frac{1}{4}''
\end{array}
$$

In a similar manner the second space between stiffeners has its rivet spacing determined. Here it is seen that the rivet spacing may start at $3\frac{1}{4}$ inches, can not exceed $3\frac{1}{2}$ inches until past the middle, and can have a few spaces at $3\frac{3}{4}$ inches at the stiffener. By Table X it is seen that 24 spaces at $3\frac{1}{4}$ inches equal $6'$-$6''$. Now $(6'7\frac{1}{4}'') - (6'$-$6'') = 1\frac{1}{4}''$ or $\frac{5}{4}$ and if one of the 24 spaces be increased $\frac{1}{4}$ inch and two of them are increased $\frac{1}{2}$ inch the entire $\frac{5}{4}$ inch will be used up and the spacing will have been completed. It is·

$$
\begin{array}{llll}
21 \text{ spaces at} & 3\frac{1}{4}'' = 5'\text{-}8\frac{1}{4}'' \\
1 \text{ ``} & \text{`` } 3\frac{1}{2}'' = 0'\text{-}3\frac{1}{2}'' \\
2 \text{ ``} & \text{`` } 3\frac{3}{4}'' = 0'\text{-}7\frac{1}{2}'' \\
\hline
& \text{Total} = 6'\text{-}7\frac{1}{4}''
\end{array}
$$

In a similar manner almost any combination can be made to fill out any dimension.

The rivets in the horizontal flange of the angle and the cover plate are, when the spacing is greater than $2\frac{5}{8}$ inches, placed opposite those in the vertical flanges as is shown in Fig. 76, since according to Table IX, Y being $(2\frac{1}{2}'' - \frac{5}{8}'') = 1\frac{7}{8}''$, no stagger is required, and where the spacing is less than $2\frac{5}{8}$ inches it is changed so as to be 3 inches or more. In such cases as this it is not necessary to give spacing in the cover plates, a note, "Spacing same as in vertical legs" or "Spacing same as in web" being all that is required.

After all the spacing in the cover plates has been determined, it may be necessary to change it slightly in order to allow for better spacing in the connection plates, but it is common practice to make the connection plates conform to the spacing in the cover plates since by so doing the additional cost of templets for the horizontal legs of angles is saved; and although a few additional templets for connection plates may be required the saving is considerable.

Cover Plates. The actual lengths required are given on the stress sheet, but when the preliminary layout is made and the material ordered, the plates are ordered longer in order that they will at least be the required length when they are on the girder. The cover plates should be stopped so that the last rivet is $1\frac{1}{2}$ inches from the end, and in the case of double-gauge angles this rivet must be on the outer gauge, see Fig. 76. The single gauge is to be recommended, providing sufficient rivets can be gotten in. At the ends of the cover plates it is not necessary to give the distances to the edges. The dimensions should go on as in Plate VIII and Fig. 76. The material notation should be put on as shown, all cover plates on both top and bottom, which are of the same section and length, being listed at the top. Any plate which is special to the bottom, is listed there.

The beginner should be careful to note that the bottom cover plate next to the flange angles does not run the entire length of the girder, and accordingly he should not run his rivet spacing in the bottom flange through to the end but should stop at the end of the cover plate. This is a common error for beginners.   .

Cross Frames. The cross frames may be detailed as shown in Plate VIII or as shown in Fig. 77. A layout of the plates must be made; the working point being taken at the intersection of the angle gauges, as in Fig. 77, or at some point which is approximately in the line connecting these points, see Plate VIII. The latter method has the advantage in that it allows the point to be so chosen that the ends of the diagonals will be about $\frac{1}{4}$ inch from both the stiffener and the top angle, thus making a smaller plate. The bevels are not stated on the diagonals since the dimensions are given directly.

The end distances should be given or, if not, a note stating their value should be on the sheet. The distance, intersection to intersection and end hole to end hole, should always be given, likewise the distance to the center of any group of holes. The rivet spacing may then be measured from these points. It was formerly customary to give the distance a, Fig. 77, but it is unnecessary and it is not now

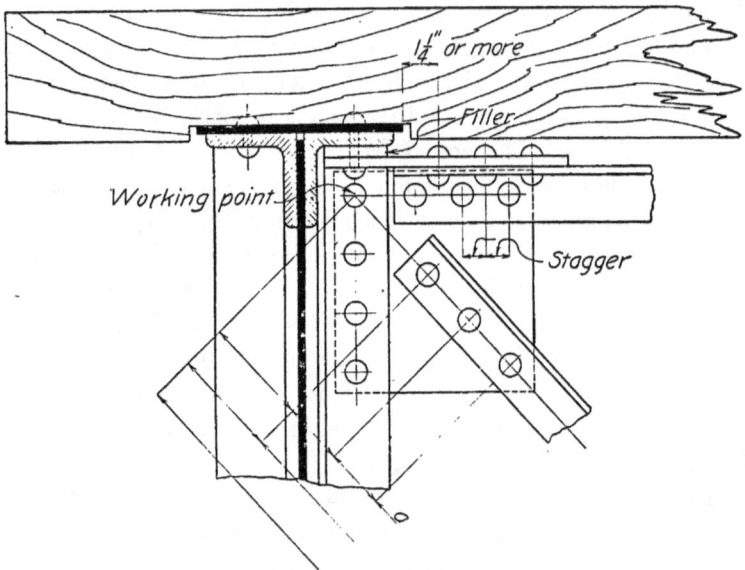

Fig. 77. Detailing of Cross Frames

put on the drawing. Attention is called to the detailing of the diagonals in C. F. 1, the center line being half way between the gauges and a rivet placed on it at the ends.

Rivet clearances should receive close attention. The first rivet in the horizontal leg of the top and bottom struts should be at least $1\frac{1}{4}$ inches away from the edge of the cover plate, and it and all others should so stagger with those in the vertical flange that the field rivets may be driven. In case the frames are as in Plate VIII, the clearances of the rivets should be looked after and the spacing in the cover plates be so arranged as to have one rivet on the gauge of the angle.

In cases where there is not a cover plate or where the cover plate is thin, the tie may, on account of its being notched $\frac{1}{2}$ inch, press down on the rivet heads of the cross frames. This may be avoided either by cutting out the tie or by placing fillers as shown in Fig. 77. Since the tie is notched at $\frac{1}{2}$ inch and the head of a $\frac{7}{8}$-inch rivet is $\frac{5}{8}$ inch, then the cover plate thickness added to that of the angle must be at least $(\frac{1}{2}+\frac{5}{8})=1\frac{1}{8}$ inches before a filler is required; and the thickness of the filler required in any case is

$$t=1\tfrac{1}{8}''-s$$

where s is the sum of the thicknesses of the cover plate and flange angle, or flange angle alone in case there is no cover plate. Of course no filler is required at the bottom. All intermediate cross frames should be alike, and the end cross frames should be like each other. In Plate VIII, the angles are $\frac{7}{8}$ inch and the first cover plate $\frac{9}{16}$ inch, the sum being $(\frac{7}{8}+\frac{9}{16})=1\frac{7}{16}$ inches, which is greater than $1\frac{1}{8}$, no filler is required.

The top angles should have their horizontal leg detailed with the cross frame. This will save many dimensions on the lateral systems when they are detailed.

Lateral Systems. The lateral systems should be detailed in place whenever possible.

All the panels of the lateral systems should be of the same length. If this is not possible, the shortest panels should be at the ends. It is seldom possible to make all the panels equal when a rolled-steel masonry plate is used. In case of the cast-steel pedestals, the dimensions of the top may be so chosen as to have all the panels of the lateral system equal. This will make the lengths of all angles with the same sized legs on connection plates equal.

The angles may be detailed as shown in Plate VIII or as in Fig. 78. In either case the distance between intersections and between end holes must be given. The rivet spacing is measured back from these reference points, being determined from the layout of the plate. The distance from the working point out to the first hole should be given. The end distance should be given or else noted somewhere else. The plate should, when a double-gauge line is used, be made to take in both rows of rivets; however, as mentioned before, the double-gauge line should only be used when unavoidable.

The working point should be in the center of the web and on the gauge line of the stiffener angle when the method used in Fig. 78 is used, except in cases where a splice is used in the center of the girder, and then the intersection or working point should be at the center of the girder, see Fig. 73. All of the methods, Plate VIII and Figs. 78a and 78b, are in common use. The author prefers those

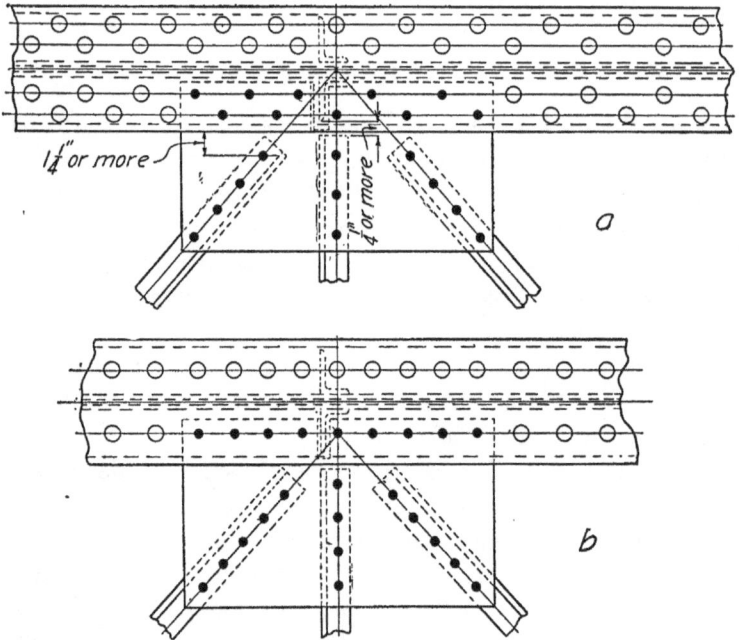

Fig. 78. Detailing of Angles in Lateral Systems

shown in 78b or Plate VIII. Sufficient clearance should be between the cross frame and stiffener, see Fig. 78b.

Each different angle, as in the case of stiffeners, should have a different mark. In such cases the mark is all that is necessary to designate another angle exactly like it, thus much repetition in detailing is avoided. The lateral systems, Plate VIII, are good examples of the efficient use of the marking system.

The size of the connection plates is determined from the layouts, the rivet spacing and clearances all being taken from the layout also.

The edge distances of the working ends and edges are shown only when greater than 1½ inches, and in some cases even then the plates are kept rectangular throughout except in the case of the smaller ones. Few sizes for many plates give evidence of good detailing, and Plate VIII exemplifies this. It might be noted that with single gauge lines in the cover plates, the plates can be detailed more economically than when the double-gauge lines are used in the angles.

A notch must be cut in the plates to allow the stiffener angle to clear. This must be carefully located and detailed for each plate where it differs in the least, and all plates to which any one notch applies should be noted directly with the detail.

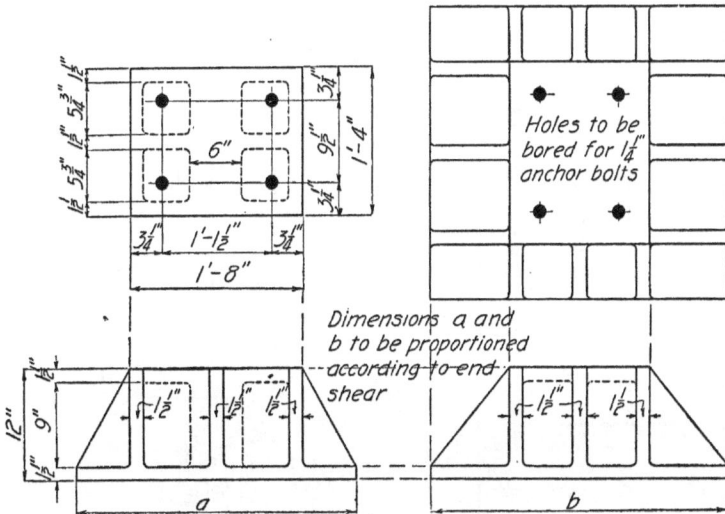

Fig. 79.   Details of Cast-Steel Bearing

Bearings. These may be as shown on Plate VIII and should be detailed in that manner, or they may consist of cast-steel pedestals with or without rollers, Figs. 79 and 80. The rollers may be either circular or segmental. In the latter case they should, in case the abutment or pier is liable to settle, have a tooth on each end of the lower plates, otherwise the movement of the girder and the movement due to the settlement of the abutment will cause the rollers to tilt over so far that they will not move back under movements due to temperature.

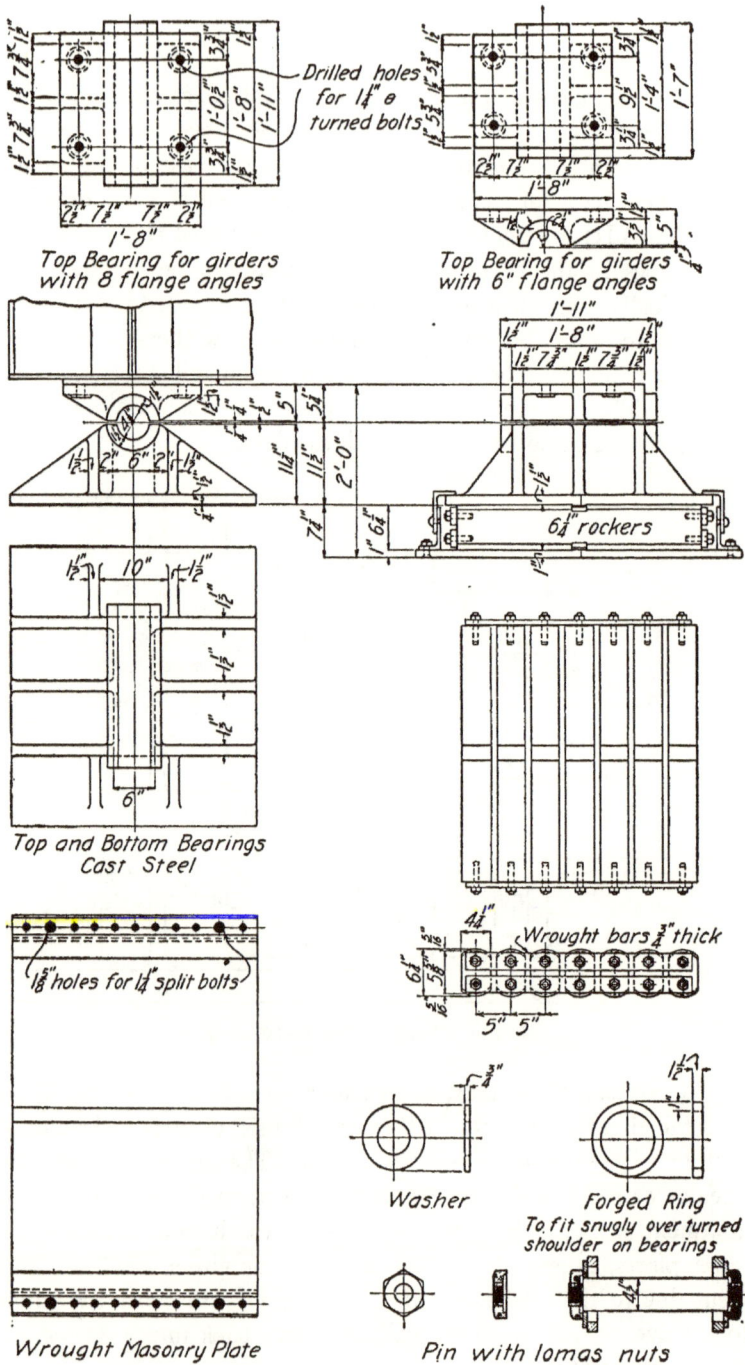

Top Bearing for girders with 8 flange angles

Drilled holes for 1¼" ⌀ turned bolts

Top Bearing for girders with 6" flange angles

6¼" rockers

Top and Bottom Bearings Cast Steel

1⅝" holes for 1¼" split bolts

Wrought bars ¾" thick

Washer

Forged Ring
To fit snugly over turned shoulder on bearings

Wrought Masonry Plate

Pin with lomas nuts

Fig. 80. Pin-Bearing Shoe and Rockers for Plate-Girder Bridges

Detailing of Compression Members. The first thing necessary is to determine the pin plates and the number of rivets required. This is done by a method already discussed.

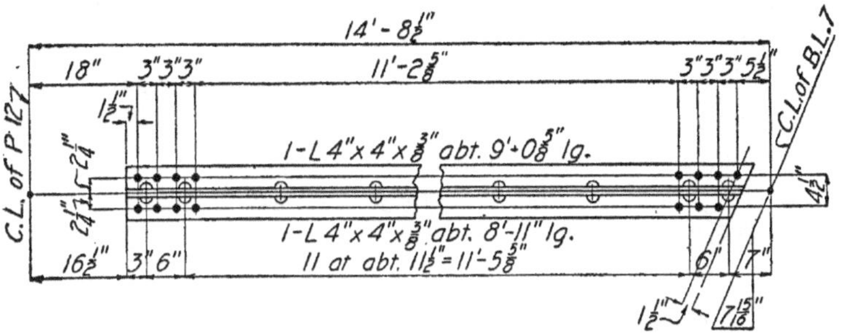

Fig. 81. Detail of a Two-Angle Compression Member

The rivet clearances and also the clearances required in order that each member may fit in with the adjacent ones in the structure, should receive the most careful consideration.

The compression members consisting of two angles riveted should be riveted together at distances throughout their length not greater than 12 inches. The clauses of the specifications relative to lattice bars, see Table XIV, and batten plates should be carefully read and followed. The dimensions necessary in compression mem-

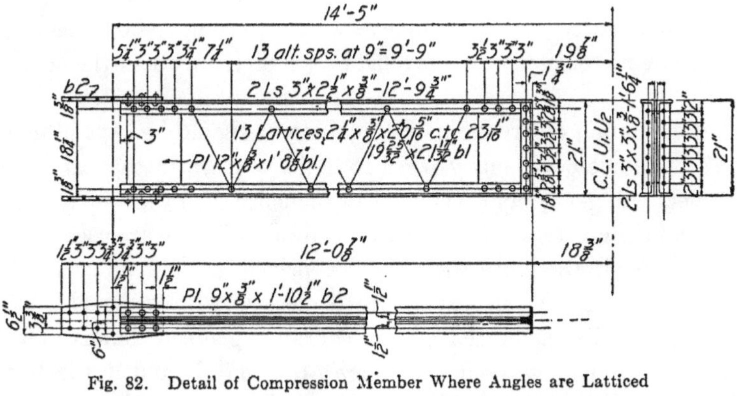

Fig. 82. Detail of Compression Member Where Angles are Latticed

bers of two angles are the same as those required in diagonals of plate girders. The spacing of the rivets which rivet the angles together need not be given but noted as "Rivets spaced about 12 inches centers." Fig. 81 is a detail of a two-angle compression member.

When two or four angles are latticed, they should have tie plates at their ends unless otherwise specified. In such cases the method of detailing is shown in Fig. 82. The ends may or may not be alike. The left-hand end is the most usual method of connection.

Compression members consisting of channels and lacing bars and tie plates are very common. Their design is given in Bridge Engineering and the clauses of the specifications cover the details. The pin plates should be on the inside, not the back, of the channel. Fig. 83 represents a typical detail of this class of member.

Compression members of cover plates and channels are used in light bridges. The detailing of such a class is shown by Fig. 84.

Heavy compression members are made up of angles and plates. The detailing of such members requires considerable care in order that the clearances may be sufficient. Fig. 85 shows a top chord section $U_0 U_2$ of a riveted railroad bridge and fairly well represents the detailing of that type of member.

Detailing of Built=Up Tension Members. The detail of these members is no different from the detailing of compression members of the same class, except that care must be taken not to reduce the section beyond the required amount, by taking out too many rivet holes. Those clauses of the specifications relating to batten or tie plates and lattice bars apply here as well as to compression members.

Built-up tension members must be symmetrical about the neutral axis.

Facilitation of Erection. In detailing, it should be kept in mind that while there are many ways to detail a piece so that the shop and field will get it right, yet some of them are such that the fabrication and the erection will be greatly facilitated if they are used. The rules to facilitate fabrication are the principles laid down in the previous pages. While experience is necessary in order that the erection will be facilitated by the correctly planned details of the draftsman, yet many points tending to this may be put in the form of rules or instructions. The following will, if attended to, tend to prevent delays and will facilitate erection.

(1) The first consideration for ease and safety in erection should be to so arrange all details, joints, and connections that a

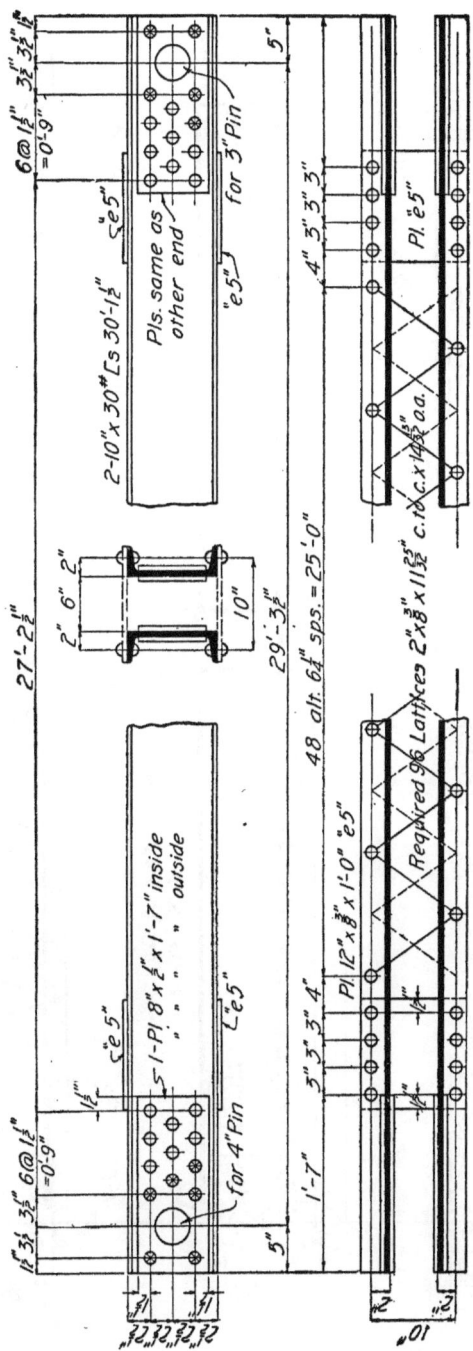

Fig. 83.   Detail of Compression Member Made up of Channels, Lacing Bars, Tie Plates

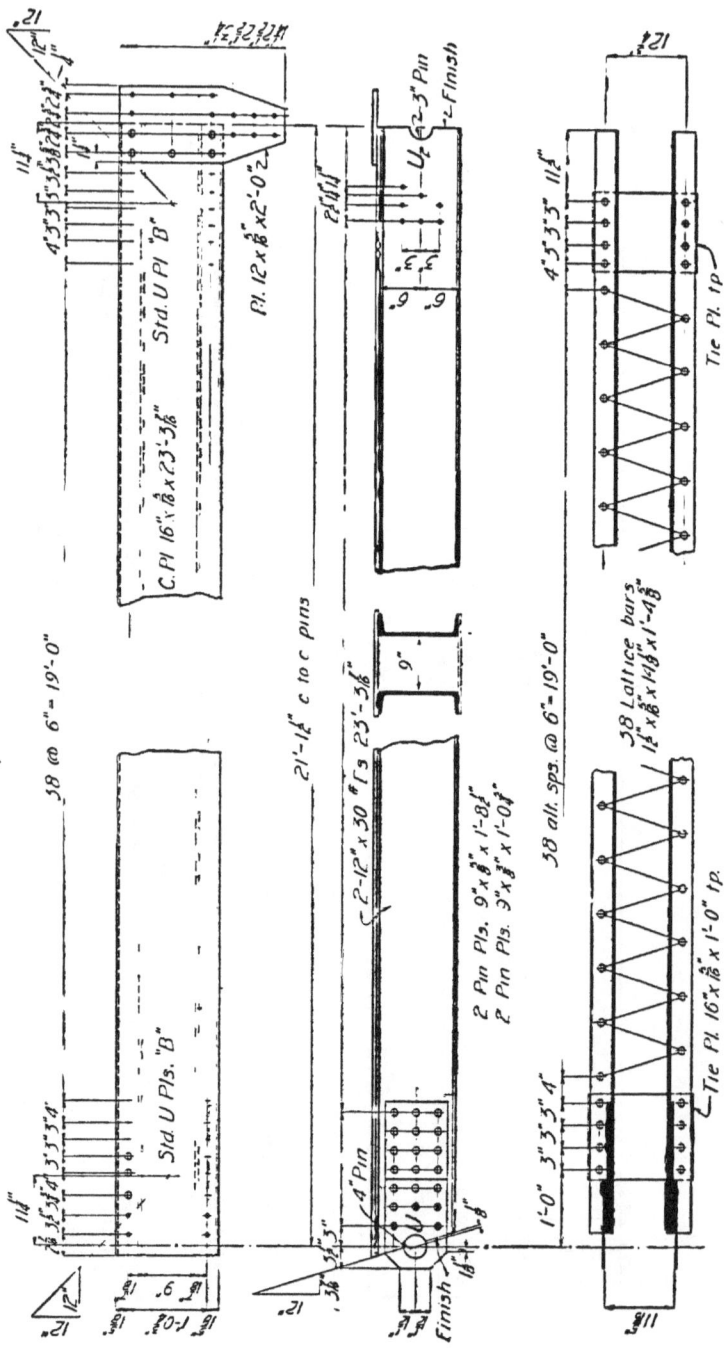

Fig. 84. Detail of Compression Member Using Cover Plates and Channels

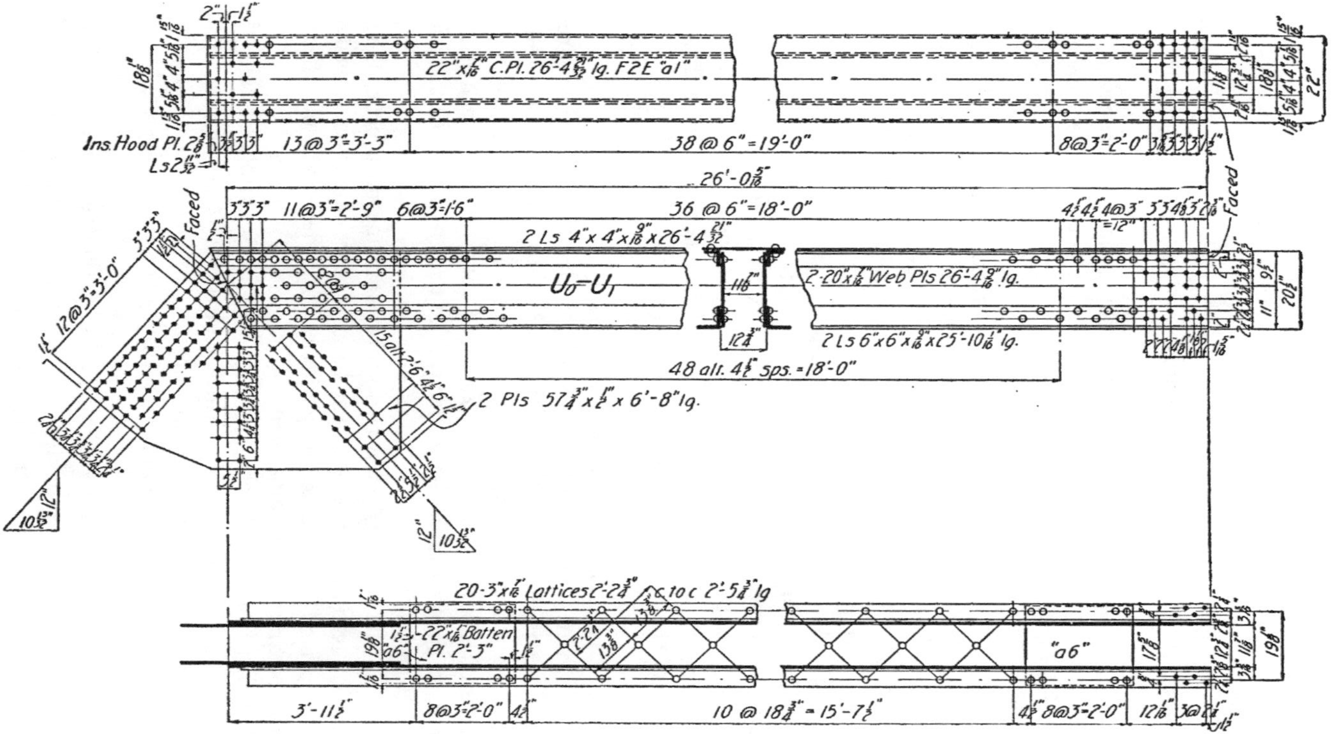

Fig. 85. Detail of a Top Chord of Riveted Railroad Bridge

structure may be connected, made self-sustaining and safe in the shortest time possible.

(2) Entering connections of any character should be avoided when possible, notably on top chords, floor beam and stringer connections, splices in girders, etc.

(3) When practicable, joints should be so arranged as to avoid having to put members together by entering them on end, as it is often impossible to get the necessary clearance in which to do this.

(4) In all through spans floor connections should be so arranged that the floor system can be put in place after the trusses or girders have been erected in their final position, and vice versa, so that the trusses or girders can be erected after the floor system has been set in place.

(5) All lateral bracing, hitch-plates, rivets in laterals, etc., should, as far as possible, be kept clear of the bottom of the ties, it being very expensive to cut out ties to clear such obstructions.

(6) Lateral plates should be shipped loose, or bolted on, so that they do not project outside of the member, whenever there is danger of them being broken off in unloading and handling.

(7) Loose fillers should be avoided. They should be tacked on with rivets, countersunk where necessary.

(8) In elevated railroad work, viaducts, and similar structures, where longitudinal girders frame into cross girders, shelf angles should be provided on the latter. In these structures the expansion joints should be so arranged that the rivets connecting the fixed span to the cross girder can be driven after the expansion span is in place.

(9) In viaducts, etc., two spans, abutting on a bent, should be so arranged that either span can be set in place entirely independent of the other. The same thing applies to girder spans of different depth resting on the same bent.

(10) Holes for anchor bolts should be so arranged that the holes in the masonry can be drilled and the bolts put in place after the structure has been erected complete. In concrete masonry they should be set very carefully according to data furnished by the Bridge Company.

(11) In structures consisting of more than one span a separate bed-plate should be provided for each shoe. This is particularly important where an old structure is to be replaced; if two shoes were put on one bed-plate or two spans connected on the same pin, it

would necessitate removing two old spans in order to erect one new one.

(12)   In pin-connected spans the sections of top chords nearest the center should be made with at least two pinholes. On skew spans the chord splices should be so located that two opposite panels can be erected without moving the traveler.

(13)   Tie plates should be kept far enough away from the joints, and enough rivets should be countersunk inside the chord, to allow of eye bars and other members being easily set in place.

(14)   Posts with channels or angles turned out and notched at the ends should, whenever possible, be avoided.

In conclusion, it may be said that the author has written this treatise with the idea of preventing the beginner from falling into the more common errors of judgment, as well as helping him to become proficient in detailing according to good common practice.

INDEX

www.ingramcontent.com/pod-product-compliance
Lightning Source LLC
Chambersburg PA
CBHW021942220326
41599CB00013BA/1654